1,000,000 Books

are available to read at

---

www.ForgottenBooks.com

Read online
Download PDF
Purchase in print

ISBN 978-1-330-31656-6
PIBN 10024737

This book is a reproduction of an important historical work. Forgotten Books uses
state-of-the-art technology to digitally reconstruct the work, preserving the original format
whilst repairing imperfections present in the aged copy. In rare cases, an imperfection in
the original, such as a blemish or missing page, may be replicated in our edition. We do,
however, repair the vast majority of imperfections successfully; any imperfections that
remain are intentionally left to preserve the state of such historical works.

Forgotten Books is a registered trademark of FB &c Ltd.
Copyright © 2018 FB &c Ltd.
FB &c Ltd, Dalton House, 60 Windsor Avenue, London, SW19 2RR.
Company number 08720141. Registered in England and Wales.

For support please visit www.forgottenbooks.com

1 MONTH OF
FREE
READING

at
www.ForgottenBooks.com

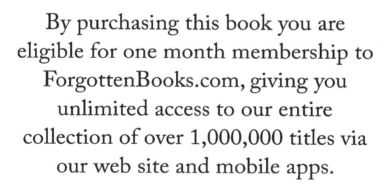

By purchasing this book you are eligible for one month membership to ForgottenBooks.com, giving you unlimited access to our entire collection of over 1,000,000 titles via our web site and mobile apps.

To claim your free month visit:
www.forgottenbooks.com/free24737

* Offer is valid for 45 days from date of purchase. Terms and conditions apply.

English
Français
Deutsche
Italiano
Español
Português

www.forgottenbooks.com

Mythology Photography **Fiction**
Fishing Christianity **Art** Cooking
Essays Buddhism Freemasonry
Medicine **Biology** Music **Ancient
Egypt** Evolution Carpentry Physics
Dance Geology **Mathematics** Fitness
Shakespeare **Folklore** Yoga Marketing
Confidence Immortality Biographies
Poetry **Psychology** Witchcraft
Electronics Chemistry History **Law**
Accounting **Philosophy** Anthropology
Alchemy Drama Quantum Mechanics
Atheism Sexual Health **Ancient History**
Entrepreneurship Languages Sport
Paleontology Needlework Islam
Metaphysics Investment Archaeology
Parenting Statistics Criminology
Motivational

Professional Paper No. 13   Series B, Descriptive Geology, 26

DEPARTMENT OF THE INTERIOR

UNITED STATES GEOLOGICAL SURVEY

CHARLES D. WALCOTT, DIRECTOR

DRAINAGE MODIFICATIONS

IN

SOUTHEASTERN OHIO AND ADJACENT PARTS OF WEST VIRGINIA AND KENTUCKY

BY

W. G. TIGHT

STANFORD LIBRARY

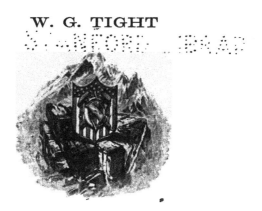

WASHINGTON

GOVERNMENT PRINTING OFFICE

1903

297920

STANFORD LIBRARY

CONTENTS.

CONTENTS.

ILLUSTRATIONS.

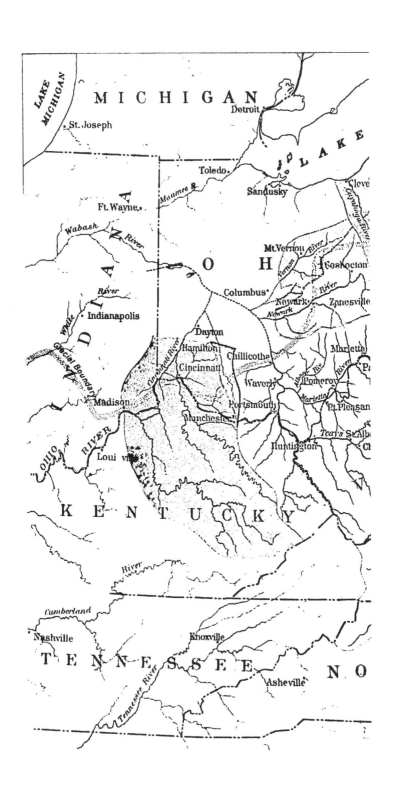

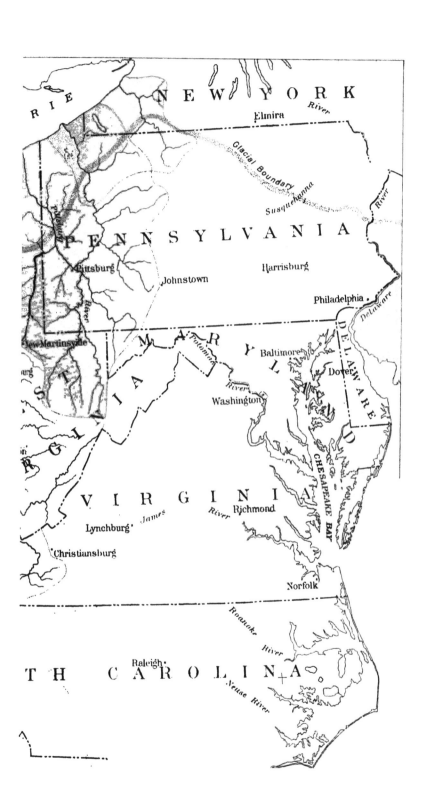

DRAINAGE MODIFICATIONS IN SOUTHEASTERN OHIO AND ADJACENT PARTS OF WEST VIRGINIA AND KENTUCKY.

By W. G. TIGHT.

INTRODUCTION.

The field work upon which this paper is based was carried on intermittently for several years. During the season of 1899 the work in Washington County, Ohio, was conducted under the direction of the Ohio State Academy of Science, the expenses being covered by a grant from the Emerson McMillin special research fund. The field work of the season of 1900 was done under the direction of Dr. T. C. Chamberlin, chief of the Division of Pleistocene Geology of the United States Geological Survey, and to him the author is greatly indebted for many valuable suggestions in relation to the work, and for an extended correspondence concerning the interpretations, during the preparation of this report.

The study of this particular region was the natural result of earlier studies of drainage modifications in Ohio, in the region more nearly adjacent to the glaciated area. Until a few years ago but very little systematic study of the drainage features of Ohio had been made. It is not intended at present to make a complete review of the early literature, but to refer only to such portions of it as bear directly on the problem in hand. On account of the lack of good maps of Ohio it has not been possible to make the maps which accompany this paper as accurate as might be desired. They have been constructed from various published maps and from personal observations in the field; and, while they are not strictly accurate, it is hoped that they will set forth the facts with reference to the drainage modifications with sufficient clearness to enable anyone to follow the features in the field, or to serve for purposes of correlation. The map of Flatwoods and Teays valleys has been constructed from four topographic sheets of the United States Geological Survey—Ironton, Kenova, Huntington, and Charleston. The author is under obligations to Mr. H. M. Wilson, geographer of the Survey, for data furnished from the unpublished map of the Kenova quadrangle. The photographs illustrating the report were all taken by the author.

The special reason for undertaking the study of this particular region and for the importance attached to the correlations is that it occupies an intermediate position between the Pennsylvania regions on the east, which include the Upper Ohio Basin, where the drainage modifications have been so suggestively studied by Messrs. Chance, Stevenson, Lesley, Wright, Lewis, White, Spencer, Foshay, Hill. Chamberlin, Leverett, and others, and on which an excellent report has been published by Dr. T. C. Chamberlin and Mr. Frank Leverett;[a] the regions of central Ohio, including the Lower Scioto Basin, on which reports have been published by the author[b] and also by Mr. Leverett;[c] and the regions farther to the west. These relations are shown on the map, Pl. I.

That the changes in the drainage within the region have been numerous was noted many times by the earlier geologists, but in all cases these changes have been considered merely as slight modifications of or deflections from the present lines, and but little attempt has been made to portray completely the drainage systems as they existed before the streams had their present courses. Within the glaciated area the drainage changes have been more extensively studied, and the causes of the modifications and deflections have been carefully worked out, with the result that the theory that the ice was a potent factor in producing such modifications in the regions over which it passed has been thoroughly established; but no explanation of the numerous changes in drainage that have been noted within this region, which lies entirely beyond the limits of the glacial invasion, has yet received general acceptance. The well-recognized factors which operate to produce drainage modifications under normal conditions of erosion have seemed insufficient to account for the phenomena here observed. It was with the hope of completing correlations of the various modifications and of determining the factors which were operative in producing them that these studies were undertaken.

At a very early stage in the work it became evident that phenomena of considerable magnitude had to be considered. Some of the old drainage channels were so situated that it was certain that the present lines are not merely deflections from them, but that they represented parts of a system of drainage which seemed to be very imperfectly related to the present system. It therefore became necessary to carry the work of collecting the data and tracing the old lines to such an extent that it was possible to map the old system. This work has now been carried so far, it is believed, that most of the important divisions of the old systems have been carefully determined. Later it may be found that in some minor details the work needs revision, and it is expected that subsequent work in the field will bring to light many new cases of modi-

---

a Am. Jour. Sci., Vol. XLVII, 1894, p. 247.
b Bull. Sci. Lab. Denison Univ., Vol. IX, pt. 2, pp. 22–32.
c Ibid., pp. 18–21.

fication which have not yet been discovered, but it is thought that these will not interfere in any way with the general correlations which have been deduced from the data already in hand.

On account of the lack of a topographic survey of Ohio, it has often been difficult to obtain satisfactory bench marks on which to base calculations of altitude. The bench marks of the railroads and of the United States Geological Survey have been utilized as far as possible. The data based on aneroid readings have been as carefully corrected as the emergencies of the case would permit. The exact elevations of the old valley floors and gradation plains are usually determined with considerable difficulty, as these are often remote from any well-established datum line. Until the topographic survey of the State is accomplished it will not be possible to present such data except in very general terms.

In determining the exact altitudes of points used as bench marks during the field work, use has been made of all available sources of information. The McFarlane Geological Railroad Guide, United States Geological Survey Bulletin No. 160, by Gannett, the river and canal surveys, the published reports of bench marks established by the United States Geological Survey, the State and private surveys, and railroad surveys have all been used. It has often happened that careful study of these various sources of information has shown the elevation of a desired station to be very doubtful. In some cases as many as five different figures are given by apparently equally good authority for the same station. In many cases the author has, by a carefully executed system of aneroid readings, been able to discriminate and determine the elevation most probably nearest the true one. In other cases, where the discrepancies were developed in the study, after the field work, and when experimental determination, even of a crude character, was impossible, various factors have been considered in determining the figure to be employed.

Through the kindness of the chief engineers of several of the railroads which traverse the region the following corrected list of elevations is offered. In view of the numerous discrepancies existing, the author does not consider it possible to state elevations in very exact terms; yet the great importance of being able to do this is recognized, since the determination of the grades of many of the valley floors is most essential, and the difficulty involved is most manifest when the grades are very low (only a few inches per mile) and the distance between stations is not very great. For these reasons, in the determination of grades, base lines as long as possible have been employed, and the elevations of stations have been taken with as much accuracy as possible. It is believed, however, that the altitudes given represent approximately the correct elevations of the old gradation plains, valley floors, benches, terraces, etc.

Elevations of stations on the Baltimore and Ohio Southwestern Railroad.

Station.	Elevation in feet.	Station.	Elevation in feet.	Station.	Elevation in feet.
Chillicothe	629.5	Hamden	713.5	Guysville	617.5
Schooley	664.5	McArthur Junction	744.5	Stewart	615.5
Vigo	634.5	Vinton	738.5	Frost	607.5
West Junction	807.5	Zalaski	713.5	Coolville	627.5
Ray	618.5	Hope Furnace	702.5	Torch	718.5
Byers Junction	644.5	Moonville	712.5	Little Hocking	630.5
Byer	648.5	Marshfield	816.5	Belpre	634.5
Richland	701.5	Athens	646.5	Parkersburg	642.5
Summit	778.5	Canaanville	628.5		

Elevations of stations on the Cincinnati, Hamilton and Dayton Railway.

Station.	Elevation in feet.	Station.	Elevation in feet.	Station.	Elevation in feet.
Chillicothe	652.2	Ironton Junction	775.6	Summit Point	758
Rupels	650.2	Wellston	734.5	Gallia	701
Richmondale	617.3	Berlin	757.2	No. 1 Tunnel	886.8
West Junction	617.3	Lowest Point	703	No. 3 Tunnel	850
Byers Junction	669.2	Rocky Hill	760	Dean	677
Coalton	710	Madison Furnace	728		
Glenroy	741.7	Cackley Swamps	718		

Elevations of stations on the Toledo and Ohio Central Railway.

Station.	Elevation in feet.	Station.	Elevation in feet.	Station.	Elevation in feet.
Palos	698.8	Hebbardsville	679.9	Langsville	579.9
Millfield	674.2	Albany	746.2	Rutland	573.7
Chauncey	657.7	Carpenter	628.3	Middleport	570.9
Armitage	653.8	Dyesville	610.4		

Elevations of stations on the Hocking Valley Railway.

Station.	Elevation in feet.	Station.	Elevation in feet.	Station.	Elevation in feet.
Logan...................	728	Bairds Summit........	796	Mills...................	564
Fivemile Run..........	708	Radcliff	645	Gallipolis.............	561
Summit	893	Hawk	639	Flood '84.............	571
Starr..................	775	Minerton	627	Addison..............	570
Swan..................	760	Niles.................	696	Cheshire	568
Summit	801	Ewington.............	677	Flood '84.............	576
Creola................	729	Vinton	606	Middleport	564
Summit	785	Glenn................	727	Pomeroy	565
Elk Fork	713	Porter 6	698	Flood '84.............	578
McArthur	730	Barren Creek	662	Low Water	571
McArthur Junction.....	738	Evergreen	690		
Summit	791	Kerr	591		

Elevations of stations on the Detroit Southern Railroad.

Station.	Elevation in feet.	Station.	Elevation in feet.	Station.	Elevation in feet.
Waverly...............	594.3	Jackson	653.1	Cornelia..............	865.5
Given.................	656.3	Summit	686.4	Denver...............	838.7
Robbins...............	669.6	Coalton	682.1	County Line..........	779.1
Beaver	668.6	Glenroy	708.4	Hills.................	652.5
Whitmans.............	699.3	Wellston	629.3	Greggs Hill...........	673.4
Cove	699.8	Hawk.................	759.4		
Simosons	670.4	Lincoln	788.6		

LOCATION OF THE AREA.

The region under consideration embraces a large section of southeastern Ohio, parts of West Virginia and Kentucky, and the southeastward extension of the Kanawha–New River into Virginia and North Carolina. It includes all the territory drained by the Ohio River and its tributaries between New Martinsville, W. Va., and Manchester, Ohio.

Before the present drainage was developed this region constituted a basin which was drained northward through the Scioto River. From the east side of Scioto River, a little above Chillicothe, the bounding watershed passed through eastern Ross County, into Hocking County, where it crossed the present North Fork of Salt Creek

a few miles above Eagle Mills; thence, passing somewhat northward around the headwaters of Raccoon Creek, it crossed the present Hocking River a few miles above
Nelsonville; thence, taking a northerly direction around the headwaters of Monday
and Sunday creeks, it crossed the Muskingum near the north line of Morgan County.
It then passed eastward in a rather tortuous line through northern Noble County
between the waters of Duck Creek and Wills Creek; thence through Monroe County
between the waters of Cranes Nest Fork of the Muskingum River and Sunfish Creek,
crossing the Ohio River a little below New Martinsville, W. Va.; thence in a rather
direct southeast line across West Virginia to the State line near Monterey; thence
southwesterly along the State line between West Virginia and Virginia to within a few
miles of the point where New River crosses the State line.   From here it turned
again toward the southeast, passing between the headwaters of the James, Roanoke,
and New rivers to the Blue Ridge mountain range; thence, passing along this mountain range toward the southwest, it crossed the Virginia–North Carolina line, passed
around the headwaters of New River, and again crossed the Virginia–North Carolina
line.   From there it crossed, in an almost direct northwest direction, the lower portion of Virginia, separating the waters of the Big Sandy from those of the Cumberland, Kentucky, and Licking.   Continuing in this general direction, bearing slightly
more to the north between the headwaters of the Little Sandy, Tygarts Creek, and
Kinniconick and the waters of the Licking, it again crossed the Ohio a few miles
above Manchester, crossing the lower waters of Ohio Brush Creek a few miles above
its mouth; thence it passed in a northerly direction, between the western tributaries
of the Scioto, to the waters of Brush Creek, and reached the Scioto again in the
vicinity of Chillicothe.   In general the area included in the watershed had the shape
of a somewhat irregular ellipse, with its long axis in a northwest-southeast direction,
one end lying near Chillicothe on the Scioto and the other high up on the eastern
plateau.

   The most noteworthy modifications of drainage are found in Washington, Morgan, Athens, Fairfield, Hocking, Vinton, Ross, Meigs, Gallia, Jackson, Pike, Scioto,
Adams, and Lawrence counties, Ohio; Pleasants, Wood, Jackson, Mason, Putnam,
Kanawha, Cabell, and Wayne counties, W. Va.; and Boyd, Greenup, and Lewis counties, Ky.   This entire region lies outside the glacial boundary except a small portion
of Fairfield and Ross counties (Pl. I).   The basins of the Muskingum above the
north line of Morgan County, of the Hocking above Logan, and of the Scioto above
Chillicothe do not properly belong within the region and will be discussed only in
their relation to it.

PRESENT DRAINAGE AND ITS RELATION TO ADJACENT DRAINAGE.

THROUGH-FLOWING STREAMS.

The streams which drain this area may be divided into two classes. The first are the through-flowing streams, which derive their water supply from outside the region and are but little affected by the precipitation within it. To this class belong the Ohio, Muskingum, and Scioto, and, to a much less extent, the Hocking.

The Ohio enters the area on its northeastern side a little below New Martinsville and passes in a general southwesterly direction to near Huntington, where its general course becomes northwesterly, which it follows to Sciotoville, when it again flows to the southwest as far as Vanceburg; here it turns again to the northwest, passing out of the area a little above Manchester.

The Muskingum enters the area at the north line of Morgan County and passes in a general southeasterly direction, but with a very crooked course, to the Ohio at Marietta.

The Hocking flows into the area a few miles above Nelsonville and passes in a general southeasterly direction to the Ohio River at Hockingport. Its larger tributaries, Monday Creek, Sunday Creek, and Federal Creek, lie within the basin.

The Scioto enters from the north at Chillicothe and passes almost directly south to the Ohio at Portsmouth.

The relation of these through-flowing streams to the normal drainage of the area forms a most interesting problem, the solution of which is attempted in the present discussion.

INDIGENOUS STREAMS.

The second class may be called the indigenous streams, as they are probably consequential and lie wholly within the bounding watershed of the basin-like area. These may be mentioned in their order, passing from New Martinsville down the Ohio, which river is, of course, the present major drainage line.

The first considerable stream is Middle Island Creek, which is on the south side of the Ohio and empties into the latter near St. Marys. Its headwaters are well up on the plateau against the side of the Monongahela Basin. Its general course is northwest, somewhat parallel to the old divide, to Middlebourne, in Tyler County; here it turns to the west and flows into the Ohio, making a very acute angle.

On the north side of the Ohio is the Little Muskingum River, which corresponds in a general way, in its relation to the Ohio, to Middle Island Creek on the south side. Its principal headwater tributaries are in Monroe County; they flow in a general southeasterly direction, turning almost squarely to the southwest when they meet the axis of the basin. The close parallelism of the Little Muskingum River with the neighboring section of the Ohio is noteworthy.

A little below the mouth of the Little Muskingum, Duck Creek enters the Ohio after traversing Noble and Washington counties. The lower course of Duck Creek is somewhat parallel to and very near the lower section of the Muskingum, and its relations to the latter stream are abnormal and very interesting.

Below the mouth of the Muskingum the next important tribuary of the Ohio is the Little Kanawha. Its headwaters extend out upon the Allegheny Plateau, even to the south of those of the Monongahela Basin, and are intimately associated with the headwaters of Elk River and Gauley River, as shown by Mr. Campbell, and referred to in his article on drainage modifications [a] and elsewhere. The general course of the Little Kanawha with its associated tributary is toward the northwest, entering the Ohio at Parkersburg.

At Little Hocking the Little Hocking River enters the Ohio on the north side. Its basin includes a large portion of western Washington County and is divided into two rather separate sections, one drained by the east branch, which heads within a few miles of Marietta and runs remarkably parallel to the Ohio for many miles, the other drained by the Little Hocking proper.

Below the Hocking, in the southern part of Athens County and in Meigs County, lies the basin of Shade River, which is divided into three sections—eastern, middle, and western. The headwaters of the eastern and middle sections extend to within a very short distance of the valley of the Hocking below Athens, and might suggest a migration of the divide toward the larger stream, but other considerations, to be brought out later, are opposed to this view. The abnormal relations between the streams in the lower section of this basin and the Ohio are noteworthy.

The tributaries of the Ohio on the southern side between Parkersburg and the Kanawha River are all small, and rarely drain more than a small part of a single county.

At Middleport, a few miles below Pomeroy, Leading Creek enters the Ohio on the north side. This stream heads in the flat lands of southern Athens County, in the vicinity of Albany, and flows in a general southerly and southeasterly direction. Its features present many abnormalities, to be mentioned later.

South of the basin of Leading Creek lies the somewhat smaller basin of Campaign Creek. This stream rises in southwestern Meigs County, and flows in a general southeasterly direction, entering the Ohio at Cheshire, some distance above Point Pleasant.

At Point Pleasant the Kanawha comes into the Ohio on the south side. The upper section of the Kanawha is known under the name of New River, and its head-

---

waters may be said to be almost on the Atlantic slope, for they lie well on the southeastern side of the Blue Ridge Mountains in North Carolina. Its course is northeast to the vicinity of Christiansburg, where it turns to the northwest, and retains this general direction to the Ohio. Some of the peculiarities of this interesting river will be more fully discussed later. It may here be stated that it is the axial stream in this basin-like area, that it is antecedent to the Appalachian elevation, and that its lower tributaries are consequential. The deflection of the lower section of the Kanawha from Teays Valley, below St. Albans, to its present position forms one of the early recognized and important changes in the area.

Some distance below Gallipolis the Ohio receives an important tributary, Raccoon Creek. The headwaters of Raccoon Creek are in southern Hocking County, and some of its branches approach very near to the valley of the Hocking. Its basin is long and slender, intercalated as it were between the valleys of the Hocking and the Scioto, and covering a considerable section of southeastern Ohio. The stream itself is very tortuous, and its abnormalities are very marked. In certain sections of its course many of its smaller tributaries are flowing in a reversed direction to that of the stream, and meet it at high angles.

A stream of somewhat similar character and peculiarities is Symmes Creek, which rises in southern Jackson County and flows through Gallia and Lawrence counties to the Ohio River opposite Huntington. Though a smaller basin than that of Raccoon Creek, it is of very similar form, and the abnormalities in its drainage are fully as marked.

At Guyandot, on the West Virginia side, the Guyandot River enters the Ohio. The lower course of the Guyandot lies within Teays Valley. At the point where it enters the valley it receives Mud River as a tributary. The latter also occupies a portion of Teays Valley west of Milton. South of Teays Valley the Guyandot and Mud rivers occupy nearly parallel basins, and with Coal River and Twelvepole Creek drain most of the territory between the Kanawha and the Big Sandy.

The Big Sandy enters the Ohio at Catlettsburg. Its headwaters lie well up on the plateau and are opposed by the headwaters of the Cumberland. Its general direction is toward the north, bearing somewhat to the west. The normal relation which the Big Sandy, Guyandot, and Coal rivers bear to the Kanawha as the axis of the basin is to be noted. This is also true of the Little Sandy and Tygarts Creek, which lie directly to the west of the Big Sandy Basin, although these latter are smaller streams and their general direction is more toward the northeast, with their headwaters extending toward those of the Licking River of Kentucky.

On the north side of the Ohio, a little above Wheelersburg, Pine Creek enters the Ohio. It drains the eastern part of Lawrence County and the western part of Scioto County; its headwaters are in Jackson County. It flows for many miles almost directly south, and then turns to the northwest, running at a very acute angle with the Ohio until it meets it. Its abnormal direction is very striking. At Sciotoville the Little Scioto River comes in on the north side also. This stream is made up of several branches which are very irregularly disposed. Its general direction is toward the south.

Below the Scioto, and within the area, the principal stream on the south side is Kinniconick River. This, with Salt Creek, drains the principal portion of Lewis County, Ky. Their headwaters are opposed by those of the Licking River, and their general direction of flow is toward the northeast, and opposite to that of the Ohio, which they meet at very sharp angles.

The other smaller streams to be noted are those tributary to the Muskingum in its lower section, which are Meigs Creek and Olive Creek on the north side and Wolf Creek on the south side. The latter is of special interest. Rising in northwestern Morgan County, it runs parallel to the Muskingum in a general southeasterly direction to the northwest corner of Washington County, where it turns abruptly to the northeast, still nearly parallel to the Muskingum, and enters the Muskingum at Beverly. The south branch of Wolf Creek rises near the mouth of the Muskingum, a little west of Marietta, and flows roughly parallel to the Muskingum, but in an exactly opposite direction, that is, to the northwest, and joins the main stream only a mile or so above Beverly. The peculiarities of this relation are at once very striking and suggestive of important modifications.

The larger tributaries of the Hocking, within the area, are Monday Creek, which rises in southern Perry County and flows southward through the eastern section of Hocking County and enters the Hocking a little below Nelsonville; and Sunday Creek, which heads in a number of branches in southern Perry County and western Morgan County, flows toward the south, and enters the Hocking near Chauncey.

The only considerable tributary on the south side of the Hocking is Margaret Creek, which rises in the flat lands around Albany in southwestern Athens County and flows northward to join the Hocking at the great loop near Athens. Its direction of flow is almost exactly opposite to that of the larger stream.

The basin of Federal Creek lies north of the Hocking, below Athens, and its drainage is extremely interesting, as the axial stream is almost in the form of a circle, and its tributaries are arranged largely radially, on the northern side. It enters the Hocking a few miles below Guysville.

In the basin of Raccoon Creek we note two tributaries on the west side. The smaller is the Elk Fork, which drains the central portion of Vinton County and flows in a general southeast direction, joining the stream a little above Radcliff; the other is the Little Raccoon, which rises in southern Vinton County and flows through the northeast corner of Jackson County to join the larger stream a little below Vinton in Gallia County.

A tributary to the Scioto on the west side is Scioto Brush Creek, which rises in numerous branches in eastern Adams County and flows almost directly east to the Scioto. Some of its southernmost headwaters rise within 2 or 3 miles of the Ohio River and flow in a northerly direction, making a circuit of more than 30 miles before they join the main stream. North of the basin of Scioto Brush Creek is the Sunfish Creek Basin, which includes the western part of Pike County, and north of this the Paint Creek Basin, which opens into the Scioto at Chillicothe.

The adjacent drainage is extremely varied in its direction and general relations. To the north is the upper section of the Scioto, Hocking, and Muskingum basins, which furnish the water supply for these through-flowing streams. To the northeast lies the upper basin of the Ohio, including the Beaver, Allegheny, and Monongahela basins, which supply the water for the through-flowing Ohio. To the east lies the Shenandoah, which touches the basin at the headwaters of the Greenbrier; and the headwaters of the James, which are also opposed to those of the Greenbrier, and reach very close to New River itself in the Great Valley. To the southeast lies the basin of the Roanoke, whose headwaters flow from the southeastern side of the Blue Ridge Mountains, and are opposed to many of the tributaries of the upper section of New River. Near the extreme headwaters of New River its smaller tributaries are opposed by the waters of Yadkin and Catawba rivers, tributaries of the Great Pedee and Santee. To the southwest several of the headwater streams of the Tennessee reach in against the divide adjacent to the upper section of New River. To the west the Kentucky and Licking river basins form the bordering areas, and to the northwest lies the smaller drainage basin of Ohio Brush Creek.

It thus appears that surrounding the area discussed and that drained by the through-flowing streams the water passes northward to the lake system, eastward to the Susquehanna and Potomac systems, southward to the Atlantic slope system, and westward to the Mississippi system.

PRE-GLACIAL DRAINAGE OF ADJACENT REGIONS.

From the extensive studies carried on in western Pennsylvania by many authors it appears practically demonstrated that the Upper Ohio Basin suffered many changes in the arrangement of its drainage lines in connection with the advance and recession of the ice during the various stages of the Glacial epoch. A summary and review of this work has been given by Dr. T. C. Chamberlin and Frank Leverett,[a] in which may be found a somewhat complete reference to the literature on the subject. The

---

a Am. Jour. Sci., 3d series, Vol. XLVII, 1894, pp. 247-283

general conclusions arrived at by these authors are that the Upper Ohio Basin was
divided into three distinct sections in pre-Glacial times, as shown on the map (Pl. I).
These are the old Upper Allegheny, the Middle Allegheny, and the Lower Allegheny
basins. The latter is included in the area under consideration. It is shown that this
Lower Allegheny system included Monongahela, Youghiogheny, Conemaugh, and
Clarion rivers as its headwater streams, and that its lower portion, which is here
termed the Pittsburg River, extended northward along the line of the reversed
Beaver and through Grand River into the Lake Erie Basin. A tributary of this
basin on the western side is called Old Upper Ohio River. Its headwaters included
those of Fishing Creek, which is now tributary to the present Ohio at New Martinsville, and from New Martinsville the general course of this stream was along that of
the present Ohio reversed to the mouth of the Beaver.

Four hypotheses are offered in interpretation of the data. In all of them it is
assumed, as supported by the facts, that the modifications which deflected all the
water of the Upper Ohio Basin over the divide at New Martinsville into the region
under discussion were inaugurated at or previous to the time of the very earliest ice
invasion.

The pre-Glacial drainage directly to the north of the basin has been the subject
of study for many years by the author, and his conclusions have been published
chiefly in the Bulletin of the Scientific Laboratories of Denison University. He has
discussed[a] the drainage modifications in the upper section of the Muskingum Basin,
and shown that the pre-Glacial course of the Tuscarawas, instead of being south
from Dresden, was toward the southwest, through an old abandoned valley, to
Newark, in Licking County, and thence toward the south and southwest to the Scioto
Basin somewhere in the vicinity of Lockbourne, south of Columbus, in Franklin
County. The general features of this course are shown on Pl. I, and the lower
section of this old river is shown on Pl. XVI, in the upper left-hand portion. One
of the tributaries of this pre-Glacial stream, for which the name Newark River is
suggested, had its headwaters near Bluerock, on the north line of Morgan County,
and flowed thence northward to the main river, receiving a considerable tributary—
the Moxahala River—at Zanesville.

Another minor tributary to the Newark River has been described by Mr. J. H.
Davis,[b] who shows that the West Fork of Jonathan Creek is a composite stream, and
that the portion west of the east line of Perry County originally flowed westward, in
pre-Glacial times, to join the Newark River somewhere in the vicinity of the old
Licking reservoir. Another somewhat larger tributary of the Newark River, which
is also contiguous to the area under discussion, has been described by the author,[c]
who has shown that the upper waters of the Hocking have been greatly modified

a Bull. Sci. Lab. Denison Univ., Vol. VIII, pt. 2, pp. 47–50.   b Idem, Vol. XI, pp. 165–178.   c Idem, Vol. IX.

and reversed; that the East Fork of Rush Creek, which heads near New Lexington, in Perry County, flowed almost directly westward past Bremen to Lancaster, where it received the waters of the upper Hocking, reversed, from a little above Nelsonville, and together the two streams discharged toward the northwest into the Newark River, somewhere in northern Perry County. These relations are expressed on the map (Pl. I).

It has recently been shown by some brief field studies that the headwaters of Clear Creek in southern Fairfield County were also tributary to this old line.

To the southwest of the Lancaster River Basin was a small basin drained by a stream for which the name Adelphi Creek is suggested. Its general relations are shown on Pl. XVI. This stream drained the western half of Hocking County. It is now a part of the basin of the North Fork of Salt Creek. The present valley of Salt Creek at Adelphi is about a mile wide and is very much obstructed with heavy glacial deposits.

A view from the cemetery hill at Adelphi toward the northwest and north shows the broad expanse of the overwash apron from the strong morainic ridge of one of the later ice invasions. While the deposits of till in the immediate valley at Adelphi seem to be of much earlier date than the morainic ridge which lies some distance northwest, the headwater streams of Salt Creek above Adelphi are all distributed over the overwash apron and the morainic ridges of the more recent glacial deposits. The broad valley formed between the moraine and the Carboniferous hills is described by G. Frederick Wright in his writings on the terminal moraine through Ohio.

To the southwest extends a long range of Carboniferous hills, over which is a more or less irregular veneering of glacial till; but it is evident that these hills offered effective resistance to the southward movement of the ice, which was not able to surmount them to any large extent.

To the east the view extends into the mouth of the valley of Laurel Creek, which is a rather broad valley bounded by rock walls having very mature contours. The direction of the valley is opposed to that of the present Salt Creek Valley, but normal to the restored drainage.

To the southeast the view is down the valley of Salt Creek, but the features of the valley indicate clearly that the view in that direction is up the stream which was instrumental in shaping the valley, for the valley diminishes in width gradually in this direction and resembles very closely the view which is obtained in looking up the Laurel Creek Valley, except that the former is somewhat larger. The till which blocks the valley at Adelphi is accumulated in hillocks rising about 125 feet above the present stream, and it appears as if these deposits had at one time formed a complete barrier across the mouth of the valley, but had more recently been cut through by the action of the stream. The till, where exposed, is of a light-yellow color and is weathered very deeply, in some places as much as 6 or 8 feet, and contains stones which are very much decayed even when obtained from several feet below the surface. Some sections of the till are bowldery, containing stones of granite, syenite, diorite, and limestone, ranging up to 3 or 4 feet in diameter. Below the weathered zone of yellow till is a light-blue till of undetermined thickness. Many of the contained stones are beautifully striated. In a few places the till shows some stratification, as though it might have been deposited in connection with a body of water.

Passing down the present stream, or up the old valley, as it were, there appear to be two systems of terraces; an upper set, which corresponds in a general way to the elevation of the till deposits that block the valley at Adelphi and that slope some-what rapidly as they pass down the stream; and a second set, which lies at a much lower level and which appears to pass westward beyond the obstructions that block the valley and to be associated with the overwash apron of the outer moraine. These two sets of terraces differ also materially in their structure. The upper terrace is made up largely of very heterogeneous material, in most cases showing considerable water action in the process of its deposition, but in others resembling closely a typical till. These terraces are much more irregular and heterogeneous in structure, and are composed of coarser material, next to the morainic deposits in the mouth of the valley, and increase in the fineness of the materials and in the amount of stratifi-cation as the distance from the till hills increases. A section about a mile from Adelphi shows, in the upper part, a weathered zone of about 6 feet of finely stratified silt; below this the material is somewhat coarser for about a foot; then comes a large lenticular mass of very coarse material about 4 feet in maximum thickness, embedded in a 15- or 20-foot bed of very fine sand; below this is a very regular belt of well-washed gravel about 2 feet thick; and below that about 20 feet of very coarse materials, showing very little stratification and yet evidently waterlaid. Scattered somewhat irregularly through the whole deposit—peppered in, as it were—are numerous stones and bowlders, a few as much as 3 feet in diameter. Some sections a little farther down the valley showed a deposit of 30 or 40 feet, composed almost entirely of very fine silt. At Haynes the valley is less than three-quarters of a mile in width, and the upper terrace deposits in this vicinity are composed entirely of fine silts and sand. In a few cases the sand is so coarse that it might almost be classed as gravel, and is as well sorted as if it had been run through a sieve. While the upper terraces are extensively preserved in the portion of the valley near Adelphi, down the stream they have been more largely removed, and in the vicinity of Haynes almost the last indication of them may be noted in a group of sand hills north of the station, almost in the mouth of the valley of a tributary creek. From the nature of this material it is easy to understand how the valley has been swept almost entirely free of the deposits. These sand hills near Haynes are very interesting remnants of the old terrace, as they have slopes of about 25° and are conical in form, rising one above the other in a progressive series to an elevation of about 200 feet above the valley floor. Below Haynes scarcely a remnant of these deposits was observed.

The terraces of the second or lower series are composed almost uniformly of well-washed coarse gravel. They slope down the stream, the material of the terrace becoming considerably finer, and the terraces themselves entirely disappearing a short distance below Haynes.

A few miles below Haynes, Salt Creek turns abruptly southwestward, while the valley continues eastward, growing rapidly smaller to South Bloomington and beyond. In this section of the valley there are deep silt deposits, which are very finely laminated, showing their deposition in slack water. These silts are leached to a great depth, so that they show no acid reaction at least 10 feet below the surface. Their material seems to be entirely of local origin, being the outwash from the neighboring Carboniferous shales and sandstones. At the point where this tributary joins Salt Creek the two valleys are about a quarter of a mile in width. From this point downstream the valley of Salt Creek gets rapidly narrower and its walls become steeper and more precipitous until it reaches its minimum width near the Vinton County line, where it is a veritable gorge. Below this point it broadens again to Eagle Mills, where it unites with a tributary of about similar size which comes in from the east. From Eagle Mills the north fork of the valley widens gradually downstream. Its floor is deeply covered with gravel deposits in the form of high terraces. Just at the point where the stream emerges from the gorge, near the Lebanon M. E. Church, these terraces are about 75 feet above the present stream; they are composed of well-rounded gravel, all of which has undoubtedly been washed from the original deposits in the upper section of Salt Creek Valley through the gorge. From the characters of the valley from Adelphi to the gorge at the Vinton County line, and the tributaries, it seems certain that in pre-Glacial times there was a col in the old divide on the county line, which will be called the Eagle Mills col, and that the drainage of the section of the valley north of this col was the reverse of that of present Salt Creek. The nature of the deposits in the upper portion of this valley, around Adelphi and to the northwest, would seem to indicate that at the time of the maximum advance of the ice its front stood across the valley at Adelphi and offered a complete obstruction to the northwestward discharge of Adelphi Creek, thus impounding the waters until they rose over the Eagle Mills col. Into these waters were discharged the great burden of glacial débris which formed the upper terraces. The coarser material would naturally be deposited nearest the front of the ice, and the material would be finer farther down the stream. In the portion of the valley beyond the main current—that is, beyond the point where the present stream turns to the southwest—the waters were almost quiet, except for the slight currents produced by the small tributaries, and in these quiet waters the heavy deposits of silt were laid down.   As the Eagle Mills col was worn away, the flood waters from the ice removed pari passu the terraces previously deposited, but on account of the smaller volume of water in these small tributaries the silts in the blind end of the valley were not so completely removed.   It would seem that this stage was followed by a recession of the ice from its position at Adelphi, which would probably mean a cessation of the supply of material, but, at least for a time, a continuation of the flood waters.   Following this there must have been a readvance of the ice, or, what seems more certain, the advance of ice of another epoch, which reached the position of the outer moraine northwest of Adelphi.   The overwash apron from this moraine extends to the deposits at Adelphi and passes down the valley of Salt Creek as a gravel train; but these gravels, which form the lower series of terraces, did not rise nearly to the height of the older series.   After the deposition of these terraces the supply of material was stopped, while the flood waters continued, and the deposits were worn out to the level of the present stream, or nearly so.   Since the falling off of the glacial floods the present stream seems to have accomplished little but the meandering of its course over the valley floor, with a certain amount of silting of its flood plain, as the stream is now on rock bottom, or nearly so, in the gorge at the Eagle Mills col.   A comparison of the deposits in the mouth of the old valley at Adelphi and the neighboring moraine beyond the apron would indicate a great difference in the age of the two moraines, and would suggest that the modifications of drainage in the Salt Creek Basin and the reversal of the old Adelphi Creek were accomplished at the time of the earliest ice invasion of this region.   This may be correlated with the Illinoian of the western region, and possibly with the Kansan, while the outer moraine would seem to belong to the Wisconsin group.

The drainage features immediately adjacent, west of the Scioto, have not been so completely worked out.   It is evident, however, that Ohio Brush Creek is a reversed stream, that the headwaters of this basin were well down toward the mouth of the present stream, and that its pre-Glacial direction of discharge was toward the northwest, as indicated on the map (Pl. I).   The drainage changes on Paint Creek are discussed by Mr. Gerard Fowke.[a]

Farther to the west of the basin the drainage features have been also somewhat fully studied by Mr. Fowke,[b] and he has shown, as indicated on the map (Pl. I), that the section of the Ohio River from the vicinity of Manchester to the mouth of the Little Miami passed back of the city of Cincinnati and joined the old Mill Creek

---

a Bull. Sci. Lab. Denison Univ., Vol. IX, pt. 1, pp. 15-24.
b Idem., Vol. XI, pp. 1-10; Ohio State Acad. Sci., Special Paper No. 3.

Valley north of the city; that the Licking River crossed the present Ohio at Cincinnati and continued northward up the Mill Creek Valley, and together these streams passed northward to the Great Miami at Hamilton. Here they received another large tributary from the southwest, made up of the section of the Ohio and its tributaries, probably to the vicinity of Madison, Ind. This stream, which is called Old Laughery, extended up the Great Miami reversed and through an old deserted valley in northwestern Hamilton County back to the Great Miami again, and thence on to Hamilton.

The conclusions of Prof. Joseph F. James,[a] of Cincinnati, with reference to the course of the Ohio up the Mill Creek Valley to Hamilton and down the Great Miami Valley, may yet prove to be an important link in the sequence of events, not only in the Cincinnati region, but in all the regions to the east.

The further continuation of the Newark River and of the northward-flowing stream at Hamilton is not so satisfactorily determined as might be desired. In his first publication[b] the author suggests reasons for believing that this drainage was discharged northwestward along the line of St. Marys reservoir and thence into eastern Indiana, where it probably turned to the southwest to join some tributary of the Mississippi Basin. Further evidence in support of this view is presented by Prof. J. A. Bownocker.[c] He has also discussed drainage modifications of the Little Miami and Great Miami systems.[d]

As the drainage basins to the southeast and south were in no way influenced by the direct action of the ice or of its deposits, the contemporaneous modifications in these regions must be worked out along other lines, and the direct correlation in point of time is attended with many difficulties; yet much admirable work has been done in this direction, although it is not possible to review the literature on the subject at this time.

GENERAL TOPOGRAPHIC FEATURES.

The general topographic features are readily divisible into two sections, viz, an upper, the New River section, composed of the mountain division and the Appalachian Valley division, and a lower, the Kanawha section, embracing the plateau region from the Appalachians to the Ohio River as one division and the lowlands north of the Ohio River as the other division. The mountain division of the New River section includes the broad expanse of mountain belt in southern Virginia and northwestern North Carolina, which extends from the Blue Ridge on the east to the Unakas on the west. Between these there is an irregular arrangement of mountain

---

[a] Jour. Cincinnati Soc. Nat. Hist., Vol. XI, pp. 96–101.
[b] Bull. Sci. Lab. Denison Univ., Vol VIII, pt. 2.
[c] Am. Geologist, Vol. XXIII. pp. 178–182; Ohio State Acad. Sci., Special Paper No. 3.
[d] Ohio State Acad. Sci., Special Paper No. 3.

groups. The eastern crests of the Blue Ridge Mountains are lower than those of the more western ranges, but the stream courses cut through the higher ranges at many points. The Appalachian Valley consists of a relatively low-lying belt, which is drained by three strikingly different systems. The northern portion is drained eastward through the gaps in the Blue Ridge made by the Roanoke and the James and, farther to the north, by the Potomac; the central portion is drained by New River, which cuts the eastward-facing escarpment of the plateau; the southern portion is drained by the tributaries of the Tennessee, and may be said to be limited by the crest line of the Blue Ridge Mountains and the eastern edge of the plateau. The floor of the valley is often broken by lesser mountain masses. The two upper divisions form but a small element in the solution of the problem in hand. They have been described in detail elsewhere by Messrs. Hayes and Campbell[a] and by Mr. Bailey Willis.[b]

The upper division of the lower, or Kanawha, section—that is, the Allegheny Plateau—is a broad table-land, sloping gently to the northwest, with its eastern edge marked by the escarpment facing the Appalachian Valley. A fine description of this table-land, with numerous illustrations, is given by Messrs. Campbell and Mendenhall in a paper entitled "Geologic section along the New and Kanawha rivers in West Virginia."[c] In this paper, referring to the gorge of the Kanawha in the plateau, the authors say:

"He [the traveler] may stand within a few hundred yards of the brink of the New River gorge, and yet be unaware that a thousand feet below him flows a mighty stream in a canyon so narrow that he can cast a pebble from the edge of the cliff almost into the stream below.

"Closer investigation shows that this even surface slopes down the stream from an elevation of 2,600 feet above sea level on the southeastern margin of the coal field to 1,000 feet at the other extremity. A still closer examination shows that it is not a continuous surface, but that at a distance from the river it is interrupted by eminences which stand distinctly above it—knobs and ridges rising out of the otherwise regular and continuous plain.

"If the traveler is a physiographer, he will see in these features an old, shallow river valley, the home of the ancestor of the present stream. * * *

"Some time during the Tertiary period of geologic history the crust of the earth remained for a long time free from oscillations, at least in the region of the coal field. The surface had previously been raised high above sea level in the interior; consequently, at the beginning of this period of quiescence erosion was very active, the streams rapidly corraded their channels nearly to baselevel, and, as time progressed, these narrow channels were changed by lateral corrasion to broad valleys as

a Geomorphology of the Southern Appalachians, by C. Willard Hayes and Marius R. Campbell: Nat. Geog. Mag., Vol VI, pp. 66–126.
b The Northern Appalachians, by Bailey Willis: National Geographic Monograph No. 6.
c Seventeenth Ann. Rept. U. S. Geol. Survey, Part II, pp 473–511.

A. VIEW ACROSS THE OHIO VALLEY AND DOWN A
SMALL RAVINE NEAR SARDIS

C. VIEW UP THE OHIO VALLEY AT LONG REACH

B. CLIFFS BORDERING THE OHIO OPPOSITE LITTLE HOCKING AT THE COL.

low as the bottoms of the original gorges. In this manner the regular surface, which now stands at altitudes varying from 1,000 to 2,600 feet, was produced near the base-level of erosion. * * *

"The topographic features of this region at the close of that interval of uninterrupted erosion were rounded and flowing, consisting of broad plains traversed by sluggish, meandering rivers, and low, rounded hills rising from the general even expanse of the broad valleys. * * *

"The termination of this condition of general quiet and uninterrupted erosion probably occurred late in Eocene time by a broad uplift, which may possibly have continued at intervals down to the present. The uplift occurred in the form of an arch with a clearly defined axial line, which passes about through Hinton, W. Va., and the movement raised the broad plain from near sea level to its present altitude of 2,600 feet. South of Hinton the uplift was less pronounced, and the surface was raised to an elevation of only 2,200 feet in the vicinity of Radford, in the valley of the upper New River. Toward the north the descent is much more rapid, the surface reaching 2,200 feet at Fayette, 1,700 feet near Montgomery, 1,100 feet at Charleston, and 1,000 feet near St. Albans.

"The first movement of this uplift caused a revival of the drainage, and the streams which had hitherto been too slugglish to corrade their channels now began actively to cut down toward sea level. * * *

"The result of this continued uplifting of the land and the downward cutting of the streams has been the production of the gorge. * * *

"Fontaine recognized the fact that the summits of the hills along New River are flat-topped, and if united would form an extended plain. He did not recognize this surface as a peneplain, but attributed it to difference in character of the rocks."

In some sections, where the hard resisting beds of the Pottsville series and the Fayette form the surface, the peneplain is well preserved, but, as these authors remark:

"Along the Kanawha River, owing to the softer character of the rocks, the peneplain was doubtless much more extensive than along New River, but the conditions which allowed of such extensive erosion also favored extensive dissection when the peneplain was uplifted, so that to-day it is difficult to find remnants of the plain which show its originally even surface. All of the ridges are terminated by sharp summits, but when the observer stands upon one of these high points the neighboring summits blend into a regular sky line, which has the appearance of an extensive plain. On account of this great regularity of summits it has been assumed that they were once points in or near the surface of a penepiain, but they are now wasted to the last degree, and may be somewhat below the altitude of the original plain."

The southwestern extension of this plateau is mapped and described by C. Willard Hayes in a paper entitled "The physiography of the Chattanooga district," [a] and is named by him the Cumberland Plateau. It is evident that the streams lying within the basin in this plateau region are consequent upon the slope of the old plain from

---

[a] Nineteenth Ann. Rept. U. S. Geol. Survey, Part II, pp. 1–58.

which the present peneplain was derived, their present courses being inherited from the streams that meandered over their broad valleys when this Tertiary plain stood near its base-level, and we can see in the extremely tortuous valleys of this region the evidence of the intrenched meanders of the older system, and occasionally, high up above the level of the present streams, may be found deserted oxbows which the revived streams have left by wearing through the necks of the upland spurs.

While the general surface of the old peneplain slopes from its maximum elevation on the edge of the eastern escarpment to the Ohio River, it does not find its lowest level along this line, but in many places crosses the Ohio River with the same uniform descent to still lower levels, and thus the plateau passes gradually into the lowlands of the Ohio division, the present drainage lines in many cases showing but slight effects on the general surface of this gradually descending plain. Along the principal divides which parallel the major water courses of the basin the elevations are often some hundreds of feet above the level of the plain nearer the main drainage lines, so that a view from the table-land along one of these higher divides presents a picture of a very gently rolling, broad basin, in which the present drainage lines are deeply impressed. The interstream tracts and broad valleys extend generally northwestward, retaining very closely relative proportions in elevation; so that, while the surface of the peneplain near the Ohio River—in the vicinity of Huntington, for example—lies about 900 feet above tide, some points in the dividing range immediately north of this in Ohio, between the waters of Symmes and Pine creeks (which are tributary to the Ohio farther down) rise considerably above this elevation. These elevated portions of the old peneplain within the lowland area therefore represent the northwestward continuation of the interfluvial tracts of the plateau division. The general features of the lowland division are not strikingly different from those of the plateau division already described. These general characteristics may be seen to good advantage from points on the divide separating the waters of the Little Muskingum from the Ohio. A view from this divide a few miles north of Sardis, looking northwestward (Pl. III, A), shows the broad basin-like surface of the plain with its gently rolling features, resembling very much the low base-level plain from which the present rough topography has been derived. Scattered over this surface are the best farming lands of the region, while the present drainage channels lie deeply intrenched in narrow valleys 100 to 150 feet, or even more, below the surface of this plain. The trees in the center of the picture mark the position of the intrenched valley, and in order to pass from the meadow in the foreground to the farmhouses in the distance it is necessary, in this case, to descend to a depth of 250 feet to the floor of the intervening valley.

On the interfluvial tracts the surface of the upland plain often rises several hundred feet above the general level. Pl. III, B, is a view near such an elevation.

B. VIEW OF THE TERTIARY PENEPLAIN OF THE LOWLAND SECTION, WITH THE SLOPE OF A MONADNOCK ON THE RIGHT.

The general level of the plain is shown on the left of the picture, while to the right is shown the character of the slope which rises to the higher level. It is against these elevated portions that the back cutting of the present erosion is most prominent, and here is offered the finest opportunity for the study of headwater phenomena.

In some sections of the basin along the eastern watershed the form and distribution of these monadnock-like elevations suggest that they are remnants of a still higher peneplain, probably of Cretaceous age, which is more perfectly preserved in the plateau regions lying farther southeast. The right edge of Pl. IV, *A*, joins the left edge of Pl. IV, *B*, and together they show some of the features indicated. The point of view is about on a level with the top of the conical hill to the left of the center of Pl. IV, *A*, and the top of the near ridge in Pl. IV, *B*. The remnant of the Tertiary peneplain furnishes the rolling lands of the upland farms on which the buildings appear. The deep, steep-sided valley at the lower left corner of Pl. IV, *A*, cut about 200 feet below the general level of the peneplain, shows the relations of the present drainage. The distant hills are 200 to 250 feet above the peneplain. Pl. IV, *B*, shows more clearly the relations of the old remnants left above the peneplain.

Beginning with the upland peneplain as a base, representing the oldest well-marked physiographic feature, we note that its surface is cut into a system of valleys which are eroded to a depth of from 150 to 250 feet below the plain. The walls of these valleys are well graded, their slopes having been reduced in most cases to very low angles. The valleys present everywhere the characteristic features of very old topography. Their floors are often wide, indicating that the streams were well graded and had accomplished considerable work in broadening their flood plains. These valleys have been the principal subject of study in this work, and their characters and distribution will be discussed more fully below. It seems certain that after the first elevation the peneplain remained in a somewhat stationary position a sufficient length of time to allow these streams to wear down their channels nearly to base-level, to extend their valleys headward, to produce an extensive distribution of low cols in the old divides, and to develop the mature topography which is associated with these old valleys. In striking contrast to the features presented by these mature forms are the valleys of the present streams. These, as already indicated, are everywhere cut into deep, narrow, V-shaped forms, varying in depth below the floor of the old valleys from a few feet in the very headwaters to 150 to 200 feet along the larger streams, or from 300 to 500 feet below the general surface of the old peneplain. The distribution of the present drainage in these deep valleys is notably different from that of the high-level valleys of the mature form, and it is evident that the modifications took place in the interval between the occupancy of the old valleys by their streams and the rejuvenation which has deeply intrenched the present drainage. The present streams in the deeper valleys do not flow on the rock floor of those valleys; they have their courses at varying elevations above the rock floor on silt or gravel fillings, and in most cases they are somewhat below the upper surface of the deposits which fill the valleys. It thus appears that a period of erosion was followed by a period in which deposition was the master feature, when the newly cut valleys were filled with river deposits varying in thickness from a few feet to 150 feet. This stage was terminated by a second period of erosion, when the streams cut deep channels in the previous deposits, leaving extensive remnants of those deposits in the form of marginal terraces along their courses; and at present, in most portions of the lowland area, the streams are still at work removing these deposits from the floors of their valleys. The most striking features, therefore, are the upland peneplain, the mature valleys of the old cycle, the deeply cut trenches of the present drainage, and the terrace filling in the bottoms of these trenches.

In some places where a present drainage line coincides with the old drainage the erosion has not completely removed the entire floor of the old valley; its fragments remain along the walls of the present stream as high-level terraces or rock platforms. Besides these prominent features, there is distributed almost everywhere over the lowland area a system of slender benches forming slight terraces on the slopes of the present valleys and of the older mature high-level valleys. These benches range in elevation from the level of the lower terraces in the present valleys to the surface of the general level of the peneplain; often the monadnocks distributed over the plain are completely encircled by one or more of these slender benches. These will receive more detailed discussion later. It is evident, however, that in point of time they are among the most recent of the topographic features.

With the exception of the modifications to be described later, the present streams are largely inherited from the drainage of the old cycle, and therefore have many features that are characteristic of mature drainage, conspicuous among which is the development of the dendritic type. On account of this dendritic distribution of the minor drainage channels and the very deep trenching, the lowland region is generally hilly and rugged, and the relief is extremely pronounced, as will appear at once from an examination of a topographic map of the Ironton region. The erosion has not, however, been so marked that the present drainage has been able to very greatly reduce the divides of the older cycle, so they persist as long and tortuous ridges with very even crest lines. In many sections these ridges are so prominent that they must be considered in the construction of highways. A road map would in many cases be almost equivalent to a relief map, as the roads are divided into two classes, known as ridge roads (Pl. XIV, E) and valley roads. These often constitute two entirely separate and parallel systems with

A VIEW FROM A HIGH RIDGE NORTH OF SARDIS, LOOKING EAST.

C. THE OHIO VALLEY, LOOKING DOWN THE RIVER FROM PARKERSBURG.

Blennerhassett Island is seen in the channel. Point of view is the same as that of *B*, opposite   The two plates together show the great bend in the
Ohio at Parkersburg

B VIEW FROM THE SAME POINT AS *A*

The left edge of this view matches the right edge of *A*

D. THE OHIO VALLEY, LOOKING UP THE RIVER FROM PARKERSBURG.

The present channel of the Little Kanawha is between the observer and the buildings in the lower right-hand corner   The trees rise from the almost vertical cliffs which face the river.

very few connections except where the valley roads pass over cols at the headwaters from one valley into another, where they intersect the ridge roads.

Wherever the valley floors of the old cycle are not occupied by the present drainage lines they form belts of flat country which are always locally known as the plains, the flats, or the flatwoods, and these names have often been helpful in locating these old valley remnants. These flat lands usually present favorable conditions for agriculture. While the deep trenching of the present drainage has made much of the region unfit for agriculture, it has here and there revealed rich mineral deposits, and has wonderfully increased the facilities for mining.

CHARACTERISTICS OF THE PRESENT RIVER VALLEYS.

THROUGH-FLOWING STREAMS.

OHIO VALLEY.

On account of its narrow, gorge-like character, the valley of the Ohio in the division of this area below New Martinsville long ago attracted the attention of geologists who visited the region. This part of the valley presents one of the longest straight stretches on the Ohio River from Pittsburg to Cairo, known as Long Reach, where there are over 16 miles of straight water (Pl. II, *C*). The bottom lands on either side of the stream are very narrow, and the valley walls are very steep. The width of the valley near Sardis is scarcely more than three-quarters of a mile, and increases gradually downstream toward Newport. To one standing on the high ground bordering the river, and but a short distance back from it, the valley of the Ohio is entirely lost to view, and one would never suspect that the great river was flowing within so short a distance, for the upland surface seems to pass from the West Virginia to the Ohio side without perceptible lowering along the line of the stream. That this part of the valley is very recent is also shown by the form of cross section of the numerous small tributaries which enter the river from both sides. These small ravines are cut down to the level of the river, and have eaten back into the uplands, forming very narrow, V-shaped gorges. Pl. II, *B*, shows the profile of one of these ravines where it enters the valley of the Ohio, and *A*, of the same plate, is a view from the Ohio side looking down one of these short ravines toward the Ohio River, the distant horizon being the level of the peneplain on the West Virginia side. The sharp angle in the graded slope of the valley wall is quite apparent.

Another noteworthy feature of this part of the valley is the sharp truncation which the river has produced on all of the bordering hills that project out as ridges against the course of the stream. Pl. II, *C*, is a view looking up the 16-mile stretch of water at Long Reach, and the steep slope of the bordering hills is quite apparent. These hills all show the gradual slope of the land during the old cycle of erosion on the face of the hill away from the river, and the steep slope of the new gorge on the face toward the river.

Pl. VI, *A*, is a view of the valley near Sistersville, W. Va., looking down the river from the foot of the bluffs on the Ohio side.

At Newport the valley is considerably wider than farther up the stream, and opposite St. Marys, on the Ohio side, there is an old channel of the Ohio which leaves the river about a mile above St. Marys and passes around a high hill, returning to the river again at Newport. This valley, the floor of which is now covered with gravel deposits, is much larger than the valley of the Ohio directly at St. Marys, and undoubtedly represents the more direct line of the stream that originally carved the old valley.

Just above St. Marys, Middle Island Creek enters the Ohio, and remnants of the old high-level floor of Middle Island Creek are well preserved. Pl. V, *B*, a view looking across the Ohio toward the mouth of Middle Island Creek and up the creek, shows the remnants of the old valley floor on both sides of the notch cut by the present stream. The old valley floor is here about 160 feet above the Ohio River (770 above tide).

Farther down the river the valley continues to increase in width, and the bordering hills become less steep and precipitous as the mouth of the Little Muskingum River is approached. These characters continue down the Ohio past Marietta, but a few miles below Marietta the valley is preceptibly narrower.

On the north side of the river a very high range of hills forms the divide between the lower waters of the Muskingum and the headwaters of the East Branch of the Little Hocking and the South Branch of Wolf Creek. These hills appear to be the northward continuation of the watershed which separates the waters of Bull Creek from those of the Little Kanawha, and the strength of this ridge seems to have been an important factor in producing the constriction of the Ohio Valley at this point.

From one of these high hills east of the Muskingum, marked Horizon Hill on the map (Pl. XI), may be obtained one of the finest views of this region anywhere in the vicinity.

On this same ridge, somewhat farther north, on a very high hill, is a red brick Catholic church with a tall spire surmounted by a golden cross, which serves as a convenient landmark for a radius of 20 to 30 miles over this section of Washington County.

Farther down the Ohio toward Parkersburg the valley of the Ohio increases rapidly in width and is from 2 to 3 miles wide, with very extensive bottoms on the Ohio side, at Belpre, near Parkersburg. The great width of the valley here is undoubtedly due to the fact that the river makes a very sharp bend and has worn

A   OLD PRE-GLACIAL VALLEY AT TORCH, LOOKING WEST FROM NEAR LITTLE HOCKING

B. REMNANT OF OLD GRADATION PLAIN AT MOUTH OF VALLEY OF MIDDLE ISLAND CREEK, LOOKING UP THE
CREEK FROM THE OPPOSITE SIDE OF THE OHIO VALLEY AND ACROSS THE OHIO BOTTOM

FROM THE ROAD, WHICH IS NEXT TO THE CLIFFS ON THE
IO SIDE

C. OHIO RIVER AND CLIFFS AT POMEROY

D. OLD VALLEY AT FLATWOODS, OHIO   *E.* OLD VALLEY NEAR DUTCH FLATS, W VA.

back the bordering hills on the outer curve. Pl. IV, *D*, is a view up the river at Parkersburg and Belpre.

Below the mouth of the Muskingum and extending down to the great bottoms at Belpre and still farther down the Ohio there are extensive gravel terraces, which represent the material carried down the Muskingum during the flood stage of the Glacial epoch. The Muskingum was one of the principal feeders to the valley of the Ohio of these glacial gravel trains. These terraces rise here to an elevation of about 110 feet above the river, or 660 feet above tide, and are extensively worked by the railroad company for gravel for ballast.

Pl. IV, *C*, is a view from Parkersburg down the Ohio. The valley is of rather uniform width from Parkersburg to Little Hocking, but it begins to narrow rapidly as it turns toward the south at the latter point, so that within a few miles it is reduced in width to about three-quarters of a mile, and at the point marked "Col" on the map (Pl. XI) it is a very precipitous gorge (Pl. II, *E*).

At Little Hocking, on the north side of the valley, at the mouth of the Little Hocking River, are extensive platforms of the old high-level valley floors, which are about 150 feet above the river, or 700 feet above tide. These platforms continue toward the west into the more complete valley floor in the vicinity of Torch.

A short distance below Little Hocking there is an old deserted waterway of the valley on the Ohio side, which is cut down nearly to the level of the present flood plain of the Ohio. This channel way, with the channel of the Ohio, incloses a considerable mass of the high table-land. It is stated by the inhabitants that the flood waters of 1884 passed through this old stream way. The sides of the channel are very precipitous, and the floor is covered with deposits of sand. It is evident from the character of the valley at the gorge that the river has here cut through a col in an old divide. This col will be called the Hockingport col. The abandoned channel was probably occupied simultaneously with the present valley of the river during the cutting of the col.

Below this col the valley of the Ohio increases gradually in width again until it reaches another maximum in the vicinity of Letart. In this stretch of the river there are several points where the valley seems to be more or less constricted, but the evidence has not been very satisfactorily worked out, so the history of this section of the valley can not be presented with as much certainty as that of others.

At Letart, and for some distance above, there are extensive bottoms on both sides of the Ohio. The great width of the valley at this point is undoubtedly due to the sharp curve in the river as it passes from its southward to its northward direction. To one standing on any of the high hills back of the river on the West Virginia side it is not difficult to see how the river, in making this great bend, has worn back the hills on the south side, giving them all a steep and truncate appearance, and why the hills on the north side, being on the inner side of the curve, and having escaped the severe erosion of the stream, are low and rolling.

North of Letart and farther down the river the valley again narrows rapidly and the walls become much more precipitous, and it is evident from the character of the table-lands that the river in this section has cut through a number of old divides, the most marked of which lies near Pomeroy (Pl. VI, *B* and *C*). Below Pomeroy the valley again broadens gradually until in the vicinity of Cheshire it is more than 2 miles wide. The bordering hills, however, are still steep and precipitous, often presenting vertical cliffs to the river front.

Opposite Cheshire, on the West Virginia side, there is a large deserted oxbow or bend of the old river. The floor of this oxbow is covered with thick deposits of sand and gravel and is known as the lower sand plains. These plains rise to an elevation of 160 to 175 feet above the present river, or about 630 feet above tide. The lower portion of these deposits, as shown by well drillings, consists largely of glacial gravels of rather uniform size, while the upper portion, amounting, as nearly as could be calculated from well drillings and sections, to 30 or 40 feet, is composed almost entirely of very fine sand, often passing into fine silts. On the northern end of this valley floor the sand has been wind-blown and is heaped up into hillocks or dunes rising 20 or 30 feet above the original plain of deposition and forming billowy ridges of considerable extent. As the gravels in this old oxbow are of glacial origin, it is evident that the river occupied this channel at some time during the deposition of the glacial gravel trains of the Ohio Valley.

While the valley at Cheshire is as wide as has been stated, the bordering hills are not so high as those in other sections of the valley, rising only about 200 feet above the river. During the flood stages the river occupies the entire width of the valley and washes the talus slopes at the base of the cliffs on both sides. During the flood of 1884 the village of Cheshire was on an island in the river, as there is a deep channel way through the terrace deposits next to the cliffs on the west side of the village.

Farther down the river the valley again narrows rapidly, and the bordering hills become higher and more precipitous. Between the mouth of the Kanawha River and Raccoon Creek (Pl. VII) the valley presents the gorge-like characters which are noticeable in the vicinity of Pomeroy and farther up the river, near New Martinsville. It is evident that here the river has again cut through several minor divides of the older cycle. There are in this vicinity a few remnants of the old gradation plains, lying at as great elevations as 200 feet above the river, or about 700 feet above tide.

Below the mouth of Raccoon Creek to the mouth of Little Guyandot Creek the valley of the Ohio is considerably wider than in the vicinity of Gallipolis, but as it

A. OHIO VALLEY AT THE COL BELOW GALLIPOLIS.

The road is built on the talus at the base of a vertical cliff   The talus is covered with big rocks like the one in the center of the view and near the roadside

B. TALUS SLOPE A HALF MILE BACK FROM THE OHIO RIVER AND ABOVE THE MOUTH OF RACCOON CREEK, NEAR THE COL BELOW GALLIPOLIS.

turns to the west below the mouth of the Little Guyandot it narrows rapidly, until at the bend near Crown City it is again a typical gorge about three-quarters of a mile in width and presenting vertical cliffs on both sides. (Pl. VIII, *A*.)

From Crown City southward the valley widens gradually and continues to increase in width until, in the vicinity of Huntington and Catlettsburg, it is from 1½ to 2 miles wide, and the bordering hills, while still often very precipitous, rise only to elevations of about 250 feet above the river.

The next marked constriction in the valley occurs but a short distance above Ashland, where the valley is again gorge-like in character and less than three-fourths of a mile wide. These changes in the width of the valley are best shown on the topographic map (Pl. IX). The map does not show the river as far as Crown City, but at the point shown at its upper limit the narrow character of the Ohio gorge may be seen, while below that point, at Guyandot and Huntington, the greater width of the valley is brought out and the constrictions near Ashland and at Ironton are also very clearly shown. The constriction at Crown City unquestionably represents the position of an old col, as does also the one at Ashland, and these will be spoken of as the Crown City and Ashland cols, respectively.

Passing farther down the river, the valley broadens again rapidly to the great bend at Wheelersburg and Sciotoville, and along this section of the valley the hills are more rolling; they rise only 200 to 250 feet above the river, and the bluffs facing the river are not so precipitous as in many other sections. Throughout this section there are numerous remnants of the old gradation plain with river deposits found as terraces at elevations of about 150 feet above the stream.

At Wheelersburg there is a well-preserved remnant of the high-level valley floor which here passes northward from the present river.

Between Sciotoville and Portsmouth the valley is again constricted, and there is another col, which will be called the Portsmouth col (Pl. XVII). The valley is not so narrow at the Portsmouth col as at many other points where it has cut through old divides, but the bordering hills are very high, rising from 350 to 400 feet above the river, and presenting very steep faces and often vertical cliffs to the river front, so that, while the valley is not so narrow, the location of a col at this point seems certain.

At Portsmouth the river bends southwestward and increases considerably in width (Pl. XIV, *E*). To an observer in the Scioto Valley the valley of the Ohio below Portsmouth seems the natural and direct continuation of the Scioto, and to the observer in the valley of the Ohio below Portsmouth, looking up the river, the Scioto Valley seems the natural and direct continuation of the Ohio Valley, as the Scioto River Valley broadens as one passes up the Scioto, while the valley of the Ohio narrows rapidly upstream above Portsmouth. These features at this point are very suggestive with reference to the direction of the old drainage lines, and indicate that the old valley was continuous between the Scioto and the portion of the Ohio below Portsmouth.

It is interesting to note that all the smaller tributaries of the Ohio farther downstream in this section enter the valley in a direction opposite to the course of the river. At Quincy, at the mouth of Kinniconick Creek, the topographic features indicate that the continuation of the valley of the Kinniconick was normally up the present Ohio Valley, or northeastward; and below the mouth of the Kinniconick the Ohio Valley begins to narrow rapidly until it reaches another marked constriction, in the vicinity of Manchester, where it is evident. from the steep sides of the valley walls and the elevation of the table-land on each side, that the river has here crossed another col, which will be known as the Manchester col. (Pl. VIII, *B*.)

Throughout the entire course of the Ohio across the area discussed the valley is marked by numerous gravel terraces, which rise to elevations varying from 60 to 120 feet above the river. These terraces are composed of glacial gravels, and represent the well-recognized gravel trains of the Ohio Valley. It is to be noted, however, that in the vicinity of the constrictions of the valley there is a marked absence of these terraces, and that they are greatest in extent below the mouths of the Muskingum, the Hocking, and the Scioto, which were the principal feeders of the glacial material. Throughout the area, also, the Ohio is now flowing about 30 to 50 feet above the rock floor. In some of the sections it is probable that the river has removed its deposits almost to the depth of the rock, while at many other points, notably at Letart and near Syracuse, the river is flowing over rock shoals. It is evident from well data that the deeper channel lies under terraces on the opposite side. From observations through the different sections of the valley it would seem as though the river is now almost in a stage of equilibrium. In some sections it is undoubtedly eroding its channel and increasing its meanders, but in other sections it is evident that at flood-water stages it is leaving over its flood-water plain extensive silt deposits, so that the vertical erosion of the stream does not appear to be very marked at the present time.

It may be stated that the characters of the valley mentioned in the region of New Martinsville would apply very generally to most of the course of the river, as the valley seems to be cut out of the old peneplain much as if it had been done by a carpenter's plane, truncating all of the bordering hills, whether they were low or high, in a very similar manner.

Another interesting feature in the relations of the minor adjacent drainage to the Ohio that may be noted in almost any section is the fact that so many of the smaller streams rise within a stone's throw of the river and make a circuitous route of several miles through deep gorges to the river, while the backward cutting of the

A. OHIO VALLEY AT CROWN CITY COL.

B. OHIO VALLEY AT THE MANCHESTER COL.

C. SLENDER WAVE TERRACE ON THE RACCOON NEAR LATHROP.

D. SLENDER WAVE TERRACE ON THE RACCOON NEAR GALLIPOLIS.

opposing ravines has been almost zero since the cutting of the great valley of the Ohio. This fact of itself seems to be conclusive evidence of the comparatively recent origin of the valley of the Ohio.

MUSKINGUM VALLEY.

Where the valley of the Muskingum crosses the north line of Morgan County it has the features of a gorge. The bordering hills have very steep slopes and often present vertical faces to the river. They rise generally from 250 to 300 feet above the river. The valley gradually broadens southward through Morgan County and reaches its maximum width in this section near Roxbury, where it bends sharply to the north and becomes rapidly narrower, while its walls grow more precipitous, until at the col near the sharp bend to the south (see map, Pl. XI) it is a narrow gorge. After passing the mouth of Meigs Creek the valley broadens again to the mouth of Wolf Creek, at Beverly, from which point it begins to narrow again on passing farther downstream, until it reaches a minimum at the point marked col on the map, a short distance above Lowell. From Lowell to its mouth it increases in size and width until, at Marietta, the valley is as large as that of the Ohio itself. (Pl. II. *D*.)

Throughout the course of the valley there are extensive gravel terraces in the broad and open portions, but these are entirely absent in the narrow section above Meigs Creek and very inconspicuous in the Lowell narrows. These terraces are the gravel trains which head far up the Tuscarawas and Licking in the morainic belts of the glaciated area.

HOCKING VALLEY.

The headwaters of the Hocking are distributed over the flat drift plain of central Fairfield County and are separated from the basin of the Scioto only by a low drift watershed. The river enters the hill country of southeastern Ohio at Lancaster, where it may first be said to be in a distinct valley. Here the valley is about a mile wide and the bordering hills are of mature form. The valley narrows gradually downstream, reaching its minimum width a few miles above Nelsonville, near Lick Run, where it crosses a very prominent north-south divide of the old cycle. This point will be known as the Lick Run col. From this point it increases gradually in width again, especially below the mouth of Monday Creek, but is again suddenly constricted a short distance below the mouth of Sunday Creek, where it crosses another col in one of the minor divides, which will be called the Sunday Creek col; and a few miles below this there is another constriction, which will be called the Sugar Creek col. The determination of the col at this point is not quite so satisfactory as at the other points mentioned.

At the great bend at Athens the valley of the Hocking is very wide, and it is evident that its great width is due to the sudden change in the direction of the stream and that the hills on the south side of the valley have been worn back, leaving extensive bottoms on the north side of the river. The features of the valley in the vicinity of Athens are shown on the map (Pl. XV), on which the broken lines represent the position of the rock walls. A short distance below Athens the valley is considerably constricted again, and the valley walls present vertical cliffs on both sides. While the valley is not here as narrow as at some other points, the presence of the old col is shown by the great persistency of the old watershed at its maximum elevation up to the very edge of the walls of the valley.

Below this col, which will be called the Athens col, the valley gradually widens and the walls become less precipitous, although they remain quite steep to the bend at Guysville. Below this point the valley gradually narrows again to the mouth of Federal Creek. Below this the narrowing is much more abrupt, and at the point marked "Col" on the map the valley is a very narrow gorge with vertical rock walls. There were here several channel ways during the cutting out of the old col by the present river. Some of these were cut nearly to the present level of the river, so that the bold rock cliffs and the numerous deep ravines present very picturesque scenery. Below this col the valley gradually broadens again and the walls become less precipitous as far down as Coolville. Between Coolville and its mouth the river again passes through narrows   That the narrows at this point are the site of an old col is not so evident as in the other cases farther up the river.

SCIOTO VALLEY.

The upper basin of the Scioto, north of a point above Chillicothe, lies wholly within the drift region, and all the drainage of the basin above this point is distributed over the surface of the great drift plain of the Scioto lobe. The streams generally occupy very shallow channel ways cut through the drift, though at some places through buried rock hills, where they have formed narrow rock gorges. These rock gorges undoubtedly determine the rate of cutting of the streams within the basin, as the rock offers much more resistance to erosion than the drift in which the streams lie throughout most of their courses. These rocky places serve as barriers to the rapid cutting of the streams. No prominent divides separate any of the tributaries of this basin, and the general features of the region are those of an extensive plain. These features are shown very clearly in the East Columbus and West Columbus topographic sheets of the United States Geological Survey.

Where the river enters the hilly district of southern Ohio, a short distance above Chillicothe, it occupies a broad valley 2 miles or more in width, with the high hills of the Waverly series rising some 200 feet above the river and forming the valley walls. These hills are well rounded and present the general features of the mature topography of the lowlands.

Southward the valley retains its width with considerable regularity; there is, however, a perceptible narrowing toward the Ohio. A few miles below Chillicothe there is an old deserted valley, behind the immediate river hills, which was at some time occupied by the river.

The southern portion of this valley is now occupied by the lower waters of Salt Creek. (See Pl. XVI.) The region has not been examined by the author, but Mr. Gerard Fowke, of Chillicothe, in a correspondence with reference to the features at this point, writes as follows:

"East of Chillicothe the river [the Scioto] flows around the base of Mount Logan on bed rock, but the drift filling still farther east is piled up more than 200 feet. This high gravel holds for several miles south and east toward Londonderry. I do not know whether the rock bed comes to the level of the stream; it is all covered up. The Salt Creek bottoms run up Little Creek, which ran in a straight course through the valley at Londonderry. There was a col on the stream above Londonderry. Salt Creek broke over this col on account of the mass of drift filling in the old channel. Londonderry stands over the mouth of Little Creek. The old Scioto flowed past Richmonddale and received the creek from the south near the county line, then swung west past Londonderry and into its present bed somewhere across the gravel deposits, and reached its present course at the first sharp bend above the mouth of Paint Creek."

Mr. Fowke further states:

"I am utterly unable to understand its [the Scioto's] present channel being so wide where the old island lies, but the hills on both sides are very steep, and it may be that the immense drift deposits are only remnants of a deposit formerly reaching bed rock, a large portion of which has been carried away by Walnut Creek since the present lines were established. Walnut Creek and the other small stream near it once came into the Scioto at the big bend above the mouth of Paint Creek. Walnut Creek now flows behind Rattlesnake Knob in a narrow gorge."

With reference to the small streams lying between the Scioto at Chillicothe and the North Fork of Salt Creek, Mr. Fowke writes:

"The four cols on these streams are clearly marked. Each stream has an easy grade from the drift land on the north between the hills, which are high on both sides at the cols. These divides were rather low, except the one farthest to the east, which latter has very precipitous hills on each side of the stream. The first three to the west afforded a channel for the water and ice to a considerable extent, but not for a great while, as there was an outlet along the glacier's front around Sugar Loaf (near the river) into the Scioto. But the last gorge, the one toward the east, had more work to do. There was a sort of reentrant curve for several miles west of Adelphi, where the accumulated water rose to the level of the col and sent a vast volume over it. There are huge sand deposits below this col all the way down the creek corresponding to those near the river. There are gullies in the sand bed 40 to 50 feet deep, where nothing is exposed from top to bottom but fine sands and silts, laminated and sloping away from the channel of main discharge, the river making its way along the hills west of Chillicothe; and this great discharge through Walnut Creek, hugging the hills to its east, made a great area of eddies and dead water in which mud settled and bergs floated, dropping bowlders of two or three tons weight."

It thus seems to be the opinion of Mr. Fowke that the river which occupied this old channelway was a northward-flowing stream, but no definite data are given as to the elevation of the rock floor of this valley. A few miles farther down the river, opposite Waverly, the valley wall of the Scioto on the east side becomes very much lower, and an extensive gradation plain of the old drainage system comes to the river at this point. While the valley at Piketon is nearly 2 miles wide, the hills on each side are high and present very steep fronts to the river, and it would seem as though the river at this point had also at some time cut off one of its older bends, as at the point below Chillicothe described by Mr. Fowke, although the data at present are not conclusive. A somewhat similar oxbow of the old valley is reported back of Lucasville, but this has not been examined by the author.

As the Ohio River is approached the valley of the Scioto becomes somewhat narrower, although its width is still very great and its walls are much higher, and, as has been already indicated in the discussion of the valley of the Ohio, the valley of the Scioto appears to be a direct continuation of the valley of the Ohio below Portsmouth.

SUMMARY.

The four valleys above described are those of the through-flowing streams, and the characteristics which they all seem to have in common are the occurrence at certain points, as indicated, of very marked constrictions in the width of the valleys, always accompanied by very steep valley walls or vertical cliffs, and in the intermediate sections a broadening of the valleys, with an accompanying reduction in the elevation of the bordering table-land and a rounding of the forms of the hills into the features of the mature topography of the lowland region.

In the case of the Ohio, Muskingum, and Scioto valleys the great gravel trains which head in the morainic tracts of the later ice invasion are marked features, and these are cut into systems of terraces, usually two or three, sometimes four, in a series.

The gravel train in the valley of the Hocking is not so extensive as in the other valleys, and the deposits seem to be of different ages, as is indicated by the amount of weathering which has taken place on the surface of the terraces. The valley seems to be swept much cleaner of these deposits, as but few of the terraces remain.

The amount of weathering on these extends to a depth of 6 to 8 feet, while the weathering on the upper terraces of the other valleys would rarely exceed 3 to 4 feet.   From the observations of the author he is of the opinion that the later gravels were not distributed down the valley of the Hocking.   However, indications of them may be discovered later.

Many of the smaller tributaries of these large rivers show interesting peculiarities in the features of their valleys, and these appear to be of considerable importance in the correlation of the abandoned high-level valley floors.

INDIGENOUS STREAMS.

MIDDLE ISLAND CREEK VALLEY.

The valley of Middle Island Creek has been examined by the author only about 15 miles back from the Ohio River.   In this section the valley is about half a mile wide, and the slopes of the hills are rather steep, though not precipitous, except at intervals.   Throughout this section of the valley there are numerous remnants of the old gradation plain occupied by the stream before the period of erosion which resulted in the cutting of the narrower trough of the present stream.   The remnants of this plain at the mouth of the valley are shown in the illustration, Pl. V, *B*.   From the topographic features in the vicinity of St. Marys, it is evident that the old gradation plain of Middle Island Creek extends along the course of the Ohio as far down as Newport, and that old Middle Island Creek emptied into the old valley of this section of the Ohio at Newport instead of above St. Marys, as at present.   Remnants of the old gradation plain still exist on the island-like hills between the old and the present channel of the Ohio, as well as on the south side of the Ohio River near St. Marys.   This gradation plain here stands about 155 feet above the river, or 770 feet above tide.

LITTLE MUSKINGUM AND DUCK CREEK VALLEYS.

These valleys have not been studied so carefully as the others, and only their general features will be referred to.   The valley of the Little Muskingum is rather narrow throughout its entire length.   It shows a marked tendency to broaden out at the points where it receives its larger tributaries.   It is cut into the floor of a broad basin-like valley of the old land surface.   One of its remarkable features is its close parallelism to the Ohio through its entire length.   The old valley of the Little Muskingum was very much larger, and had reached a more advanced stage of planation than the streamway which was later occupied by the Ohio.   A view looking northward from this divide across the Little Muskingum country is in very striking contrast to one looking southward across the Ohio.

The valley of Duck Creek resembles that of the Little Muskingum. In its lower part the valley is much broader and the hills are more rounded than in its middle and upper sections. This lower course has the appearance of recent occupancy by a stream larger than that which originally cut the valley. This fact, together with some of the features observable farther up the valley, indicates that there have been several modifications of the stream courses, but they have not been fully worked out, and are noted by question marks on the map, which may serve as suggestions for later investigators.

WOLF CREEK VALLEY.

Wolf Creek is a tributary to the Muskingum at Beverly. It heads in northern Morgan County on the divide that was crossed by the Muskingum when it broke over into this basin. It flows southward many miles closely parallel to the Muskingum, much as the Little Muskingum parallels the Ohio. Its upper valley is very narrow and deep. Southward it broadens gradually to the point where it turns eastward, when it narrows rapidly to the point marked "Col" on the map (Pl. XI), a few miles above its mouth. Near the mouth of the valley just above the junction of its south fork there is an old deserted oxbow of considerable interest. This oxbow seems to have been cut off at the time the flood waters cut out the "Col" above. The valley is narrow at the cut-off of this oxbow. The hill which occupies the center of the oxbow rises almost as high as the surrounding general surface. Below the mouth of the south fork the valley is very broad and the hills are more rolling. This valley does not seem to have ever been cut down to the level of the deep channel of the Muskingum. It seems as though the limestone stratum which forms the floor of the valley at its mouth had prevented the valley from becoming well graded to the level of the deeper channels of the larger streams.

The valley of the South Fork of Wolf Creek is markedly different from that of the main creek. Throughout most of its length it is comparatively broad and open and bounded by gently rolling hills, though at some exceptional places the walls are rather steep. In the upper portion the contrast with the headwater features of the main stream are most striking. The country around the headwaters is rather flat or gently rolling, with very deep soils. Many of the smaller tributaries rise in extensive swamp areas. These swamp areas often lie on the divide which separates the waters of Wolf Creek from those of the Little Hocking. The slope of this divide on the north side, which is drained by the tributaries of Wolf Creek, is much less dissected than the south slope, which is drained by the tributaries of the Little Hocking.

LITTLE KANAWHA VALLEY.

The valley of the Little Kanawha has been examined by the author for but a short distance above Parkersburg. Its features are, however, similar to those of all the valleys of the plateau region. Immediately at Parkersburg the present stream has cut off a rocky projection from the west side of the valley, while the older channel-way lies farther east under the eastern part of the city. On the floor of this old valley are extensive deposits of silt, but the elevation of the rock floor was not obtained.

LITTLE HOCKING VALLEY.

The Little Hocking Valley is divided into two main branches, which are very similar in character and present no exceptional or unusual features. They are rather narrow with moderately steep valley sides. Everywhere are present the marks of the recent rejuvenation. The valley of the East Fork occupies much the broader depression in the old land surface. Several of its tributaries on the north side, like the headwaters of the South Fork of Wolf Creek, rise in the flat tracts on this divide. The tributaries on the south side of the East Fork are all short, as the East Fork, like the Little Muskingum, parallels the Ohio throughout its entire length and is separated from it by a high ridge but a few miles wide.

The gradation plain of the old stream is well preserved near the mouth of the valley and is continuous with the old valley floor, which passes eastward from the Ohio Valley at this point. This plain stands about 150 feet above the Ohio, or 770 feet above tide.

FEDERAL CREEK VALLEY.

The valley of Federal Creek is rather deep and narrow in its lower portion, but in the section around Amesville it is much broader. All the tributaries on the north side occupy rather broad valleys. The effects of the rejuvenation of the drainage of the region, which are so marked throughout most of the basin, are less apparent in the Federal Creek Valley than almost anywhere else, and at the same time the slender benches mentioned in connection with the discussion of general topographic features of the lowland area are better preserved in this valley and in neighboring portions of the Hocking than in almost any other part of the basin. It is the opinion of the author that the present stream has cut through an old divide a short distance below Amesville, but the topographic features indicating a col at this point are not well marked, and the location of the col here is based more upon the distribution of the old valley floors to the south of the basin and the general features of the tributary streams than upon the features of the topography immediately adjacent to the col.

SUNDAY AND MONDAY CREEK VALLEYS.

The valleys of Sunday and Monday creeks, the two large tribuaries of the Hocking above Athens, present rather normal features. In both valleys numerous remnants of the old gradation plain are preserved. In the Sunday Creek Valley, at Chauncey, the old valley floor now stands at an elevation of about 725 feet above tide.

SHADE RIVER VALLEY.

Only the lower portion of the Shade River Basin has been examined by the author, and two marked constrictions in this portion of the valley have been observed, one on the East Fork of Shade River, a few miles above its mouth, the other on the main stream a short distance below the mouth of the East Fork. In both of these places the valleys are very narrow, less than a quarter of a mile wide, and their walls are very steep and precipitous, while the other portions of the valleys of Shade River and its tributaries present normal features.

There are evidences of some minor modifications on the Middle and West forks, but these have not been carefully worked out as yet.

South of Chester the old valley of the Shade passes along the line of Rays Run, reversed, and here the old gradation plain lies at an elevation of about 680 feet above tide.

LEADING CREEK VALLEY.

But a limited examination has been made of the upper section of the Leading Creek Valley, and sufficient data are not at hand to describe in detail the features of this portion of the basin. In the middle section of the valley two notable constrictions are observed, one in the vicinity of Carpenter and the other near Langsville. The location of old cols at these points is suggested, but the data are somewhat incomplete, and the cols are only tentatively placed as indicated on the map. In the lower portion of Leading Creek Valley there are many interesting features, to properly present which would require a carefully detailed survey. The old gradation plain is represented by many remnants, which stand at about 160 feet above the Ohio, or 680 feet above tide, and the present stream has cut a very deep, steep-sided valley into the floor of the old valley. Thomas Fork, a small tributary entering Leading Creek but a short distance above its mouth, also presents many interesting features. There is a marked constriction in its valley at Coalton, and it is very evident that this stream has here crossed an old section of the divide in its present course. This stream occupies a very deep gorge, which is only about a half a mile back from the Ohio north of Pomeroy, while its headwaters rise on the old valley floor at Flatwoods, between Pomeroy and Chester. It is very evident that the gorge of Thomas Fork was cut out by torrential waters and not by the small stream which now occupies it. It is suggested that the flood waters of the Ohio at a stage prior to the deposition of the glacial gravels may have occupied the gorge of Thomas Fork simultaneously with its present channel at Pomeroy, and that the Pomeroy channel finally obtained control of the full volume of water.

KANAWHA VALLEY.

The features of the Kanawha Valley above Charleston have been admirably described and illustrated by Messrs. Campbell and Mendenhall in a paper on the "Geologic section along the New and Kanawha rivers in West Virginia," [a] already referred to, and are also most admirably shown on the Charleston topographic sheet of the United States Geological Survey, a portion of which is reproduced in general features on Pl. IX, so that further description will not be given.

From Charleston to the Ohio at Point Pleasant the characters of the valley are not strikingly different from those in the plateau section above, but there is a notable constriction of the valley in the vicinity of Winfield that has been interpreted by others, as well as the author, to indicate the location at this point of an old col, which will be called the Winfield col. The gradation plains in the vicinity of St. Albans will be further discussed in connection with the description of Teays Valley.

There is another notable constriction of the valley at Point Pleasant, where the floor of the old valley extends eastward around the hills immediately bordering the river. At this col the valley is about three-fourths of a mile wide, and the bordering hills are very precipitous and rise to elevations of 250 feet on both sides of the river. This might be called the Point Pleasant col.

RACCOON CREEK VALLEY.

The valley of Raccoon Creek is very tortuous, and presents a great variety of features, as do also many of its smaller tributaries. The most important constrictions of the valley occur at Moonville and near Minerton. At both of these places the stream occupies a very rough and rugged section of country. All of the minor streams in the immediate vicinity of these constrictions have cut very deep narrow gorges, and the regions are as rough as any found within the borders of the State of Ohio.

At Moonville the valley is less than an eighth of a mile wide, and the walls rise in vertical cliffs to an elevation of 225 feet above the stream. The valley is here traversed by the Baltimore and Ohio Southwestern Railroad, and much difficulty was experienced in the location of the roadbed, as it was necessary to make numerous rock cuts, a few short tunnels, and several bridges in order to get through the narrows of this gorge. The features of the gorge may be readily observed from the windows of the passing train. This gorge marks the position of what will be called the Moonville col. On the south side of the valley, near Zalaski, the wall of the valley is low, and there is a continuation of the old valley directly southward.

---

[a] Seventeenth Ann Rept U S Geol. Survey, Pt. II, pp. 473-511.

At Minerton the valley is not as narrow as at Moonville, but the surrounding table-land is fully as high, and the evidence of the former existence of a col at this point is very complete. This will be called the Minerton col.

From the Moonville col southward down the creek the valley broadens somewhat rapidly and the bordering hills become more rounded until the mouth of Long Run is reached, where the width of the valley increases abruptly and the older features are much more manifest, and these conditions continue down the stream to the mouth of Elk Fork, where the valley seems to turn up this fork and continue with similar characters nearly to Vinton station. Below the mouth of Elk Fork the older features of the valley continue downstream to the great loop near Radcliff, where the valley is constricted very suddenly, and it grows narrower and more precipitous from this point down to the Minerton col. The broad features of the valley of the Elk Fork do not continue up that stream from Vinton to McArthur, but turn to the west along one of its smaller tributaries to McArthur Junction, while between Vinton and McArthur Junction the stream is in a very narrow gorge, which undoubtedly represents the position of another old col, called the McArthur col.

Down the Raccoon from the Minerton col the valley becomes wider and less precipitous, although the walls are still very steep and the region is very rough and rugged for many miles.

In the vicinity of Rio Grande the valley of the Raccoon is much broader than in any other portion of the stream's course, and the neighboring region is less rugged, the hills being very much lower and having gentler slopes. Many remnants of the old gradation plains are well preserved in this section.

Between Rio Grande and the mouth of the stream there are three well-marked valleys which lead off from the valley of the Raccoon, one toward Rodney, another, some miles farther down, toward the south, in the direction of the waters of Symmes Creek, and another at Northuptown, which passes eastward toward the Ohio in the vicinity of Gallipolis.

From Rio Grande to the mouth of the creek the valley becomes perceptibly narrower, indicating very strongly that this section of the Raccoon is reversed, a fact that is also indicated by the levels of the gradation plains, which show a gradual fall from the Ohio River to Rio Grande.

The valley of the Little Raccoon was not examined in detail, but it is evident that it is made up of sections of several independent old valleys. Only one col was definitely located on this stream, a short distance below the mouth of Dickson Fork. There are probably two or three others farther up the stream between this Dickson Fork col and Hamden.

SYMMES CREEK VALLEY.

Passing up Symmes Creek from the Ohio opposite Huntington, the valley of the creek, which is less than half a mile wide at the river, becomes a very narrow gorge, deeply cut below the high table-land in the vicinity of Marion. North of this gorge, which undoubtedly represents the position of an old col on this stream, and which is called the Marion col, the valley of the stream broadens rapidly to the junction of Sand Fork. The valley of Sand Fork is rather broad, and the stream in its lower course meanders over an extensive swampy plain half a mile or more in width. While the valley of Symmes Creek from the Marion col, as well as the northward continuation of the valley represented by Long Creek, is rather rough and precipitous in character, it is quite comparable in size to that of Sand Fork, and it is very suggestive that the principal erosion of the two valleys was produced by streams of about equal volume, but that Long Creek Valley has been recently enlarged and reversed by the present waters of Symmes Creek.

From the junction of Sand Fork with Symmes Creek a broad, open, flat country extends across the divide to the Raccoon, while farther upstream, near Evans Mill, the creek runs in a very narrow, deep, precipitous gorge. This gorge undoubtedly represents the position of another eroded col. From the Evans Mill col northward up Grassy Fork the valley broadens gradually and its walls become less precipitous, but up Black Fork another deep gorge is encountered but a short distance from the junction of these two streams. The valley of Grassy Fork, as the name indicates, is a broad, open, fertile valley for a number of miles, and on the left branch of Grassy Fork opens out into the Cackley Swamps east of Camba, while the right branch passes through a narrow gap a short distance above Madison Furnace, and its headwaters extend into the flat meadow lands south of Rocky Hill. Besides the gorge on the Black Fork, near its junction with Grassy Fork, there is still another gorge 4 or 5 miles farther up the stream. These undoubtedly represent the former positions of old cols which have been cut through by the present streams.

BIG SANDY VALLEY.

The valley of the Big Sandy has been examined by the author only for a distance of 12 or 15 miles from the Ohio. Throughout the section examined there are many well-preserved remnants of the old valley floor, those near the mouth of the river standing at about 180 feet above stream, or about 670 feet above tide. The present stream has cut a narrow gorge in the old floor, while the bordering hills are rather steep and rise to elevations of from 700 to 900 feet above tide. Within the more recent valley there are considerable deposits of silt and sand, which have been recut to a very large extent by the present stream. The silt terraces are about 130 feet above the river, and correspond in elevation to the gravel terraces that fill the deep gorge of the Ohio near Catlettsburg, and are probably of the same age, having been deposited at about the same stage of water level.   The old gradation plain remnants rise gradually up the Big Sandy Valley.

PINE CREEK VALLEY.

The valley of Pine Creek has not been carefully studied by the author, but has been observed at several widely separated points.   The lower portion of the valley, which is roughly parallel to the Ohio, has all the characteristics of valleys of the older cycle, as well as the accompanying rejuvenation presented by the deeper trenching of the present creek.   The valley, however, narrows rapidly northward upstream, which suggests that within a short distance it might pass into a gorge having the character of a col.   This suggestion is strengthened by the fact that at a point 15 or 20 miles farther upstream the creek flows in a rather wide and open valley, bordered by low hills, but the exact location of the connection between this upper larger valley and the old drainage line to the west was not determined, yet the possible locations of such a connection are suggested by question marks on the map (Pl. XVII).   These features, taken in connection with the abnormal relation of the present stream to both the old and present drainage, make it quite evident that the present stream is composed of several sections of an older drainage channel, and suggest an interesting field study.

LITTLE SCIOTO VALLEY.

The lower portion of the Little Scioto River is cut into the floor of an old valley, and its trough is about 150 feet below the plane of the old floor, while the hills bordering the old valley rise about 250 feet above the old floor.   The upper waters of the Little Scioto have not been critically examined, but that there are numerous minor modifications is very apparent from the abnormal arrangement of the drainage. The close relation to the upper waters of Pine Creek is manifest.

SUMMARY.

From the foregoing discussion of the valleys of the smaller streams that lie wholly within the basin it is evident that they possess very many features in common. They are marked by characters somewhat similar to those mentioned in connection with the larger streams, in that throughout their courses there are stretches where the valleys are wide and present features of maturity, and other shorter intervals where they are narrow and precipitous and bear all the marks of very recent erosion. Many of the smaller streams that have not been examined undoubtedly will show similar features, which will be brought out by future field work, but it is thought that enough of the characters have been determined to furnish a good foundation for an interpretation of the major topographic features. On many of the other smaller streams represented on the maps accompanying this paper some observations have been made, and wherever the characters of the valleys have been indicated they are very certainly known, but a description of each individual valley would be merely repetition, for they all bear the impress of an older mature cycle of erosion, the present streams being deeply intrenched in old valleys; and although in most cases no remnants of the older valley floors remain, the rejuvenation is clearly marked in the sharp angle and change of grade of the valley walls.

In the above discussion the location of the old cols on the various streams has been made largely on the basis of the characters of the valleys, yet at the same time the characters of the divides which separate the different basins, and of the spurs which run out from the main divides, together with the general surface slopes of the old peneplain, have also been important factors in their location; but as these features are not apparent in the immediate study of the valleys they have not been mentioned in the discussion. In this connection it may be said that in the vicinity of the cols the divides on opposite sides of the valley usually retain their full elevation up to the very edges of the valleys, while in the broader portions of the valleys the lateral divides approach the streams in series of ridges, which gradually decrease in elevation to the immediate valley walls. While it seems evident that the cols crossed by the streams were the lowest in the divides which imponded the waters, at the same time it also seems evident that they were, as a rule, very narrow gaps; and consequently the streams, in cutting down these gaps to greater depths and in broadening their valleys, cut back laterally against the divides, thus producing the features noted in the descriptions of the valleys at these points.

OLD HIGH-LEVEL VALLEYS.

GENERAL DESCRIPTION.

The old deserted high-level valleys possess many features in common. The fact that they are not now traversed by the streams that produced them or by other streams indicates that they must occupy portions of the present divides, for whenever a river valley is deserted it at once develops in some part of its course a water parting, although this may be represented for some time only by a lagoon or a swampy area. As the adjacent drainage lines are cut deeper, the old valley floor will be left higher and the divide will become more marked. This is true even as to a minor modification, such as an oxbow cut-off, but in this case the divide is usually very near the upper end of the deserted portion of the valley, the lower portion having the longer drainage line, yet these old high-level valleys all occupy positions in the present divides. It is a well-observed fact, however, that the divides of which they form a part decrease in elevation, as they approach these old valley segments, by a rather gradual series of descending ridges, until the edge of the old valley wall is reached. It is also true that the major divides of the neighboring region run parallel to the direction of these old valley floors, and it is only the spurs from these major divides that approach the sides of the old valleys and in which the floors of the valley form a part of the present divides.

Among the most common and striking characteristics of these old deserted valleys are the well-graded valley walls which border them and the very mature drainage shown by the valleys. The depth to which the old valleys are cut below the surface of the old Tertiary peneplain varies considerably, depending in each case on the location of the valley in the basin, but in general it may be said that they are usually from about 150 to 250 feet deep. The elevation of their floors above present drainage lines is also variable, but in the portions of the valleys bordering the through-flowing streams it is often as much as 150 feet above the streams. That the streams occupying these old valleys were well graded is indicated by the fact that in most places, especially on the larger streams, the valley floors are flat-bottomed, forming a graded plain often half a mile or more in width.

Another important feature associated with these old valleys is the great depth of the residual soils on the adjacent slopes and in the valleys of the smaller streams which have not been subjected to strong erosive action. Upon the floors of most of these valleys may be found deposits made by the streams. These often consist of gravel beds, sometimes of rather coarse material, but more frequently of finer sands and silts. In numerous sections of these deposits the shingling of the gravels is well marked, indicating beyond peradventure the direction of flow of the streams which deposited them. In most cases the deposits laid down by the old streams are overlain by fine silts and clays, which are undoubtedly of later age, for there are in many places evidences of buried soils below these silts and clays, and this soil zone is usually well marked between the upper deposits and the lower.

Another marked feature which is characteristic of the deposits on the floors of all of these valleys is the great depth to which they have been leached. In several cases this leaching has been observed to extend to a depth of 30 or even 40 feet. The flat lands over which these clays and silts are distributed are often known as the "crayfish" flats, and also as the "white-clay" districts. As a general rule they are very poor farming lands, and it is evident that their sterility is due to the lack of lime salts, for in some places where land plaster has been applied these poor lands have been converted into very good farms. It has been suggested that the lack of lime in these soils is due to some conditions existing during the period of the deposition of the clay; but, after an examination of a large number of sections, it is the opinion of the author that the poor quality of the soil—arising from the lack of lime—is due to the very extensive leaching which these deposits have suffered. In some sections, where exposures are obtained at a depth of between 20 to 30 feet below the surface, the lower beds of sand and gravel are often very thoroughly cemented with lime and iron salts, and it seems certain that these minerals have been derived from the upper layers by descending percolating waters. An attempt has been made to use the amount of this leaching as a time measure—to determine by it the length of time that these deposits have been exposed, but only the most general results have been obtained, and these indicate a very long period. The conditions for thorough leaching are extremely favorable, on account of the underlying porous sands and gravels, which would afford the very best of drainage to the upper silt deposits. In most cases these old valley floors are now very much dissected by the minor drainage lines which cross them, and one traveling lengthwise of the valley finds its floor often rough and hilly, so that its features may be well observed only by ascending to the top of the table-land, from which point the sharp erosion on the valley floors is lost in perspective and the broad features of the valley stand out in general outline.

Besides these positive characters, there are some negative ones which are also worthy of mention on account of the possible relation which the old drainage sustained to the drainage of the Glacial epoch. The possible occurrence of glacial deposits on the floors of the high-level valleys was constantly kept in mind and diligent search was made for them, but none were found at the higher levels, though some were seen in old oxbow cut-offs that lie at much lower elevations along the lines of some of the larger through-flowing streams—as, for example, in the old valleys at Newport and opposite Cheshire, on the Ohio, and in the old valley at Londonderry, on the Scioto.

Another bit of negative evidence is the lack of any indication of rejuvenation on the floors of any of these old valleys before the deposition of the sands and silts, this indicating beyond question that the first change in the history of the older cycle of erosion was the deposition of the sands and silts.

GENERAL DISTRIBUTION.

These old high-level valleys are distributed very irregularly over the entire area. Some, as the California Valley, the Flatwoods Valley, and Teays Valley, lie adjacent to the large through-flowing streams; other smaller remnants are distributed very generally among the headwater streams, extending even far up onto the plateau region. It is noticeable, however, that they are very much more numerous in the northeastern half of the basin, or, to speak more accurately, northeast of the major divide which separates the waters of the Kanawha and Elk River from those of the Little Kanawha, and which, continued westward, crosses the Kanawha at Winfield, the Ohio at Crown City, and Symmes Creek at Marion, as shown on the map (Pl. XIII).   For convenience of reference this divide will be called the Winfield-Crown City-Marion divide.

The most important modifications south of this divide are the Flatwoods and Teays valleys, although there are undoubtedly others that have not yet been discovered.

SPECIAL DESCRIPTIONS.

In discussing the characters of the separate remnants of these old valleys it is proposed to give to each section a name which will serve as a ready means of reference, and in the selection of these names it has seemed best to give to the valley the name borne by the town which is located upon the old valley floor or the local name by which the region is known.   It seems best, also, to discuss these different valley remnants in the order in which they are locally grouped, and to follow out in a general way the connections that exist between them.   They are grouped into two sections, according to their general distribution, a northern and a southern, and this order will be followed in the descriptions.

TEAYS AND FLATWOODS VALLEYS.

In the southern section the most important of these deserted valleys is the well-known Teays Valley.   Its relation to the Kanawha River was early recognized, and it has been described by many authors, notably by Prof. I. C. White, of Morgantown, W. Va., and Prof. G. Frederick Wright, of Oberlin, Ohio.   Since the completion of the topographic sheets of the United States Geological Survey including this region it is possible to give a much more detailed description of the characters of this valley and to present its features more clearly.   This the author has attempted to do on the map (Pl. IX), on which contour lines of 500, 700, 900, and 1,100 feet elevation are shown, and on which a system of coloring is employed to bring out the features of the bordering table-land and the general characters of the valley.

The other important old drainage way in the southern section is the Flatwoods Valley, back of Ashland, Ky., and opposite Ironton, Ohio; and as Teays and Flatwoods valleys unquestionably form parts of the same system of drainage, the two will be discussed together, as the features of the Flatwoods Valley are also shown on the topographic map (Pl. IX), taken from the Ironton and Kenova sheets of the United States Geological Survey.   It is to be observed from this map that the surface of the old peneplain in the vicinity of St. Albans and Charleston now stands at about 1,000 feet above tide, and that it slopes gradually northwestward, its elevation being about 900 feet in the vicinity of Huntington and about 800 feet

at Ironton, or that it has, in general, a slope of about 200 feet in about 50 miles, or of about 4 feet to the mile.   The Kanawha River at St. Albans is 560 feet above tide, and at Ironton the Ohio is 483 feet above tide, so that there is but 77 feet fall in the present drainage between St. Albans and Ironton.   The distance by water being 120 miles, the fall is about 7.6 inches to the mile, or, if the distance in a direct line were taken to be 50 miles, the fall would be approximately 15 inches to the mile. The old gradation plains near St. Albans are now approximately 675 feet above tide; the rock floor of the valley, as near as can be estimated, is about 662 feet above tide near Cades, in Teays Valley; at Hurricane, about 665 feet above tide; at Barboursville, at the junction of Mud and Guyandot rivers, about 653 feet above tide. Immediately back of Ashland the old river gravels are resting on the original valley floor at 645 feet above tide, while at Advance, opposite Ironton, in the old valley, the rock floor stands at an elevation of about 650 feet above tide.   Where the old valley meets the Ohio below Ironton the old gravels show the old valley floor to be about 645 feet above tide.   It thus appears that the grade of the old valley floor as it now stands is about 7.2 inches to the mile between St. Albans and Ironton.   If the grades represented by the difference in elevation of the present drainage at St. Albans and Ironton (77 feet) and that of the old valley (30 feet) were compared with the grade of the peneplain it would appear (as the present grade when measured in a direct line is nearly double the old grade) that there has been an elevation of the upper section of the region sufficient to produce the increased grade; but the fact that the present drainage, when measured in its circuitous route of 120 miles, has almost exactly the grade which is now presented by the old valley floor, would indicate that the position of the region has remained almost constant, for it seems reasonable to suppose that the volume of water in the Kanawha has been almost constant, and that the grade which it would establish in the same system of rocks would be about the same.   As the old valley has the same grade as the present stream, it appears that something other than differential movement must account for the difference in the grades as indicated, when measured in a direct line.

Since the abandonment of the old valley the present drainage has lowered its channel about 100 feet at St. Albans and about 160 feet at Ironton—the deeper cutting of the intermediate stage being neglected for the moment—making a difference of about 60 feet.   The direct line being about 50 miles and the line of present drainage about 120 miles, the establishment of the grade of 7.6 inches per mile over the longer route would more than account for the observed difference in cutting at the two points.   To enable the river to establish the same grade over the longer route would necessitate the elevation of the entire region at least that amount, or the discovery of a shorter outlet to the sea by the lower waters of the stream, with a resulting increase of the grade in that section to an amount which would permit the cutting observed in the upper section. If the difference was produced by elevation, it seems evident that it was of an epeirogenic type and included the entire basin, as the relations of these grades do not seem to indicate any differential movement, certainly none of sufficient magnitude to have produced the deflection of the drainage from the old line.

In comparing the width of the valley in different sections it is to be noticed that there are one or two places where the old valley seems to be somewhat constricted, notably at Barboursville, in Teays Valley, and in the section between Ashland and Ironton, in the Flatwoods Valley. In the latter case the valley wall extends out from the south toward the present Ohio River, and the old valley between these southern hills and the hills next to the river is rather narrow. However, even at these constricted places the valley floor is still nearly three-quarters of a mile wide, while the average width of the valley floor throughout the entire length of the valley would equal about a mile. The contrast presented by this part of the valley of the Ohio with that about a mile above Ashland is very striking, for at the point last named the Ohio Valley is scarcely more than half a mile wide and very steep bluffs stand on both sides of the river, while in the constrictions mentioned in the old valley the bordering hills are low and well graded to angles of from about 20° to 23°. The difference in the age of the two valleys is thus strikingly shown, and indicates clearly that the Ohio could not have occupied its present position in its valley above Ashland for a period nearly so long as that during which the stream that cut the old Flatwoods Valley held its course. This difference becomes still more marked when we take into consideration the fact that the stream that occupied the Flatwoods Valley was probably of very nearly the size of the present Kanawha River, which, of course, is much smaller than the Ohio.

The old river gravels on the floor of Teays Valley and of the Flatwoods Valley have been often described, and their origin has undoubtedly been correctly attributed to the headwaters of the Kanawha in the crystalline area of the Blue Ridge Mountains. Sections of these quartz gravels were examined in many places, and one interesting deposit near the lower end of the Flatwoods Valley, on a small tributary of Pond Run, showed a great number of quartz bowlders measuring 8 to 10 inches in diameter. These were resting on an old blue muck clay bed, which in turn was underlain by a residual soil passing rather abruptly into the decayed underlying rock. In most of the sections, however, the quartz gravels are much finer, and as a rule are composed of pebbles from the size of beans to the size of one's thumb. The fact was noted that the gravels were very uniform in size in each exposure, showing that the sorting action was very perfect. One exposure in a small ravine back of Ashland, resting directly upon the decomposed shales, was composed almost entirely of stones about the size of one's fist. Throughout both the Flatwoods and Teays valleys these gravels are a certain source of water supply, and wherever the present drainage is below the level of the gravels their outcrop is marked by a line of springs, while the wells drilled in these valleys always obtain water on striking the gravels.

Almost everywhere the gravels are overlain by a deposit of fine silt. Where best preserved, this deposit of silt is seen to extend clear across the valley, but in most sections of the valley it is deeply cut out by the present streams, and undoubtedly extensive areas of the old silts have been thus removed, so that it is often difficult to determine just how thick the original deposits were in any particular part of the old valley floor. At Advance, in the Flatwoods Valley, opposite Ashland, the wells penetrate about 18 to 25 feet of silt before they strike the gravel, and in several sections, shown by the ravines which cut the floor, about the same thickness of silt was observed. The silts are much thicker in the upper portion of the valley, near Hurricane and Scotts station, than they are in any other part. Here it was estimated that the silts attain a thickness of about 80 feet. They extend very generally into the smaller tributary valleys which come in along the sides of the main valley, and often produce slack-water conditions in the smaller streams which occupy these valleys, and, in some cases, even swampy conditions. From the nature of these silt deposits it seems evident that they were laid down in a body of water which stretched entirely across the valley. In no case do they exhibit marks of strong current action; on the contrary, the delicate lamination which the deposits show in some sections indicate that they were laid down in relatively quiet but at the same time not stagnant waters. Old valley soils occur below these silts quite commonly in sections that are favorably located. In several places the silts were observed at levels considerably below the quartz gravels, which are presumed to rest upon the old valley floor, and appeared very much as though they might be in place, but a more critical examination showed that these silts have very generally crept down the slopes of the deeper drainage channels which dissect the floor of the old valley, and that, in places, this creeping has been in large masses, so that the characteristics of the original silt deposits have been preserved and yet the silts have thus been carried below the level of their original plain of deposition.

As already indicated the old floor of the valley is very generally deeply cut by the present drainage channels, and has at some time been cut even deeper than at present. This is very clearly shown in the vicinity of Barboursville, where Mud River joins the Guyandot. Just north of the village, at the point marked "A" on the map (fig. 1), there is a narrow ridge, whose major axis extends north and south, and which thus occupies a position right across the valley. From its peculiar position and its association with the deep erosion of the Guyandot and Mud rivers, lying, respectively, to the west and the east, it offers a sort of natural fortification commanding the approach from either side, and it has many historic associations in connection with the civil war.   The marks of the old earthworks are still visible.

The ridge is cut through by the Chesapeake and Ohio Railroad and an excellent section is thus afforded.   About 20 feet above the railroad is a stratum of characteristic gravel overlain by about 8 feet of silt and soil.   One standing on the top of the ridge may readily observe that it is a remnant of the old valley floor, the level of which can be easily recognized by numerous other remnants, both up and down the valley (marked "VF." on the map, fig. 1).   At the very base of this ridge, on the west side, flows Guyandot River.   It is separated from the north wall of the valley by a very narrow rock gorge, through which flows Mud River.   On the east side

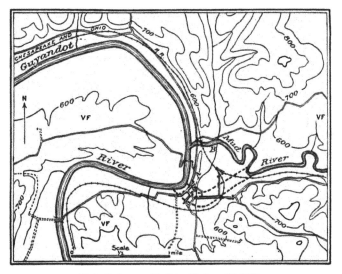

Fig. 1.—Drainage modifications at Barboursville, W. Va.

there is a low-lying area, which has undoubtedly been eroded by an old oxbow bend of Mud River.   At the south end the ridge drops down rather abruptly about 75 feet to the level of the village and is separated by this flat from the south wall of the valley.   Numerous wells in the village show that this flat, which separates the ridge from the south wall of the valley, is an old silt-filled channel extending from Mud River to the Guyandot, along the line indicated on the map by dotted lines.   As the Guyandot emerges from the hills on the south side of the valley it first turns eastward and then swings around to the west in a great curve.   Where it passes the end of this old silt-filled channel it has exposed a vertical section of the old silts about 60

A.

B.

C.

SILT CLIFFS AT BARBOURSVILLE ON THE GUYANDOTTE RIVER.

feet in thickness. Here the bottom of the river is on the silt and the water is 15 to 20 feet deep. This gives an observed thickness of 75 to 80 feet. As stated, these silts here stand in vertical cliffs resembling those which are shown by the loess. The character of the exposure is shown in Pl. X.

Just above the rock gorge which separates the ridge from the north wall of the valley, Mud River is flowing in a channel cut in the silt deposits which fill its old, deeper valley. In the gorge the river is flowing over the rock. At the point marked "B" on the map (fig. 1) the river has a fall of about 5 feet. This fall is maintained by the presence of a thin band of very compact slaty shale. From the falls to the Guyandot is about 250 yards, and in this distance Mud River descends about 5 feet over a series of much softer shales. It is evident that at the time of formation of the falls at the mouth of Mud River they were about 10 feet high. Since that time Mud River has cut back only about 250 yards. The cutting of the river through the ridge above the falls exposes about 40 feet of rock. The work involved in cutting out this 40 feet of somewhat decayed rock seems to be comparable to the work of the river in eroding the soft silts above by the lateral cutting in the low area just east of the ridge. The rock barrier probably caused the meander just above. Similarly the Guyandot found the work of excavating the silts from the old channel so much easier than working against the rock walls that it made its eastward detour on that account. From the above facts it appears that, following the desertion of Teays Valley by the Kanawha and the consequent development of Mud River on the old valley floor, there was an extended cycle of erosion, during which Mud River cut the deep channel under Barboursville and at the same time it had its point of confluence with the Guyandot near where the latter enters the valley, next to the south wall.

It would seem that the force of the Guyandot waters not arrested by the confluence with the Kanawha was directed in a straight line across the old valley. Together the two streams eroded the valley in this vicinity considerably below the present level. Following this was a period of extensive silting, and the deeply eroded channel was filled up to the level of the flat on which the village stands. This silting so obstructed the lower course of Mud River that its waters were turned over the low col in the ridge at the gorge. Then followed a period of erosion. The Guyandot, working only on the old silts, has been able to lower its channel faster than Mud River, which had to excavate the rocks in the gorge. Since the two streams encountered the hard layer of shale at their present point of confluence, the Guyandot has lowered its channel about 10 feet, while the falls of Mud River in the same time have receded about 250 yards. The rate of cutting of the Guyandot is probably limited by that of the Ohio below.

The relations of the present drainage to the features of the old valley are shown also in connection with the Hurricane Creek drainage. This stream rises in the

areas south of the valley, comes into the valley opposite Hurricane, passes out of
the valley to the north about a mile from the town, and crosses the hilly region to the
north, entering the Kanawha at about 4 miles below Winfield.   Where this stream
cuts across the valley it is in a trench about 35 feet below the surface of the valley
and about 40 feet below the gravel layer which marks the old valley floor.   In
several places the gravels and silts above have crept down into the trough of the
Hurricane.   An examination of the smaller tributary valleys, down the Hurricane a
mile or two, at once reveals the fact that the talus slopes bordering the almost perpen-
dicular walls, which are very characteristic of all of the drainage in this section
of the Hurricane Valley, are supplying material to the streams at present at a
much greater rate than they are able to remove it.   This is true even of Hurricane
Creek itself.

In this section also there are numerous hills entirely surrounded with vertical
cliffs and separated from the neighboring ridges by deep gorges of considerable
width, which are not now occupied by any water courses whatever.   These features
are also quite common on the section of Poplar Fork of Hurricane between its
junction with the main stream and the valley above, for this stream also rises on the
south side of the valley, and, like Hurricane, traverses the valley from south to
north.   It is very evident that the gorges associated with the Hurricane and Poplar
Fork have been occupied by much larger streams than those now occupying them,
and also that the present streams are running at elevations considerably above
the rock floors of the valleys, while the streams on the south side of the valley
do not show the same characteristics.   These features of the topography in this
section of the Hurricane and Poplar Fork valleys suggest at once that they may
have been produced at the time of the transfer of the waters of the Kanawha from
Teays Valley to the present course of the river across the Winfield col, and that at
the time of the transfer a portion of the waters passed around and through these
gorges on Hurricane and Poplar Fork, and then down the Hurricane to the valley
below, until the Winfield col was cut to such a depth that it drew off the upper
waters from these smaller valleys, leaving many of the deserted channel ways which
are found there at present, while the deeper cutting of the gorge of the Kanawha
and the Ohio carried the floor of these drainage lines far below the present level,
and that, with the filling of the Ohio Valley, these floors were again silted up, and
this very largely from the material supplied from the steep slopes of the gorges
previously cut.

The physiographic history of Teays Valley seems to be, briefly, as follows,
beginning with the old Tertiary peneplain as a base: First there was a very long
cycle of erosion which produced the old topographic features of the valley and its

well-graded floor, which must have stood near base-level for a considerable length of time. During the latter stages of this cycle deposits of sand and gravel were laid down along the course of the stream, while the floor of the valley was marked with numerous lagoons and the deep-valley soils which are now seen in sections in the old muck beds and buried vegetation associated with the gravel deposits. Following this there was a period of slack water, during which the valley was occupied by a very sluggish stream, in which the deep silt deposits were laid down to a depth of from about 20 to 80 feet. As the much greater thickness of these silts in the upper section of the valley in the vicinity of Hurricane and Scotts station can hardly be attributed to the difference produced by subsequent erosion, it would seem that the deposition must have been much greater at this point, and that here the upper waters of the Kanawha may have first encountered the slack-water conditions which prevailed in the lower section of its valley, and so would here have dropped the major part of its load of sediments. Following this stage of slack-water conditions and the deposition of the silt was the abandonment of the valley by the river and its meander northward across the Winfield col around the silt obstruction. Next came a long period of erosion in which the local drainage was intrenched to a depth of at least 150 feet below the old floor in the vicinity of Barboursville and corresponding depths over the completely graded courses of all of the streams, even to the smaller headwaters. Following this was the period of slack-water conditions in which were deposited the great beds of silt which filled the old valley of Mud River at Barboursville and produced a corresponding aggradation of all of the drainage lines. Then followed another period of erosion, in which the streams cut out these aggradation plains to a considerable depth, amounting to about 60 feet at the mouth of Mud River. Associated with this rejuvenation are a number of minor modifications in the positions of the streams and their deflection over minor rock spurs which were buried under the silts, of which the most noted example observed by the writer was the deflection of Mud River at Barboursville.

The general relations which the Teays and the Flatwoods valleys bear to the other drainage features are indicated on maps, Pls. XIII and XVII.

Farther down the Ohio, below Ironton, there are several well-preserved remnants of the floor of the old valley which show from their elevations a gradual descent toward the northwest. These are undoubtedly to be correlated with the Teays and Flatwoods valley floors, as they fall into close accord with the grade of the upper section and carry deposits of similar character. These are found as far north as Wheelersburg.

CALIFORNIA VALLEY.

North of Wheelersburg and near Stockdale, in Scioto County, and extending into Pike and Jackson counties, is the old California Valley, which has been described by the author and Mr. Frank Leverett,[a] and correlated, with considerable confidence, with the Flatwoods and Teays valleys, on the basis of the grade of the old valley floor and the deposits found upon it, which consist of gravel beds of quartz pebbles overlain by silt deposits.   The southern portion of this valley is very deeply cut by the Little Scioto River and its tributaries, but the northern portion, in the vicinity of California Flats, and the area extending thence northward to the Scioto at Waverly, are very perfectly preserved.   Its features may easily be observed from the train on the Ohio Southern Railroad at Beaver.   Between the Ohio at Wheelersburg and the Scioto at Waverly this floor has at present at least five water partings across it, which is suggestive of the great length of time which has elapsed since its abandonment and the establishment of the present drainage.   The elevation of the valley floor where it is cut by the Scioto Valley is, as near as can be estimated, about 600 feet above tide. This would make a grade of about 9 inches to the mile between St. Albans, W. Va., and this point, a distance of about 100 miles.   As indicated in the articles above referred to, the topographic features of the California Valley are quite identical with those of the Teays and Flatwoods valleys, and the history of the cycles through which this region has passed is also exactly similar to that of the other valley.

LAYMAN, BARLOW, AND FLEMING VALLEYS.

North of the Winfield-Crown City-Marion divide the remnants of the old valley floors are much more numerous than in the section to the south, as previously mentioned, and their discussion will be taken up in connection with their relation to the present drainage lines, and their correlations suggested in order.   In the divide separating the waters of Wolf Creek from those of the Little Hocking, in western Washington County, Ohio, shown on the map (Pl. XI), there are three well-marked abandoned valley floors.   These are near Layman, Barlow, and Fleming.   The valley floor at Barlow (Pl. XII, *C*) was described by Dr. S. P. Hildreth,[b] as early as 1838, in his report, as follows:

"On Mr. Lawton's farm, in Barlow Township, Washington County, in the midst of the marl region, is a locality of fossil fresh-water shells of the genus *Unio*.   They are imbedded in coarse sand or gravel cemented by ferruginous matter.   The spot in which they are found has once evidently been the bed of an ancient lake or pond. It is now a beautiful valley of a mile or more in width by 4 miles in length, surrounded by low hills.   On the south side a small branch stream drains the superfluous water into the Little Hockhocking.   In digging wells for domestic use in this tract, beds of sand, gravel, and plastic clay are passed to the depth of 30 feet, containing imbedded branches of trees, leaves, and fragments of wood, of recent and living species.   Similar valleys and levels are found in the uplands of the western

---

a Bull. Sci. Lab Denison Univ., Vol. IX, pt 2.   b First Ann. Rept. Geol. Survey Ohio, 1838, p 50.

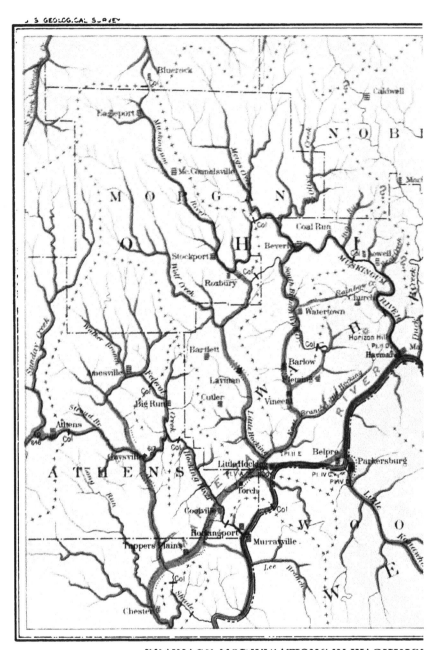

DRAINAGE MODIFICATIONS IN WASHING'

B

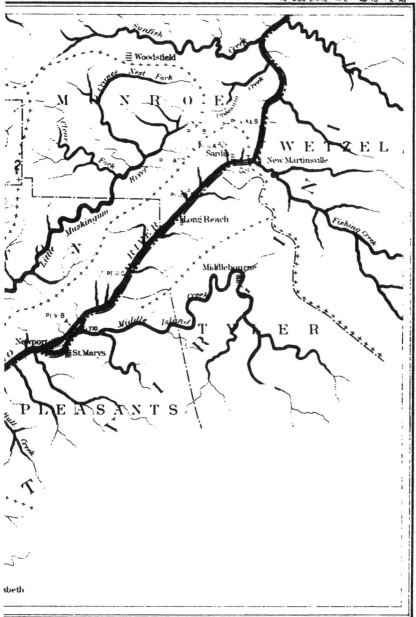

JULIUS BIEN & CO LITH N.Y

'NTY AND ADJACENT COUNTIES, OHIO

r

15   20 miles

part of the county, lying between the headwaters of the creeks, and are a kind of table-land. From the frequency of these flat lands between the headwaters of the Little Hockhocking and the south branch of Wolf Creek it is possible that at some remote period the waters of Wolf Creek were discharged into the Ohio River instead of the Muskingum. This opinion is strengthened from the fact that the head branches of the South Fork now rise within 2 miles of the Ohio, and run northerly parallel with and opposite to the course of the Muskingum for 12 miles, and join that river 20 miles from its mouth. The remains of its ancient beds would form pools and ponds of standing water furnishing fit residences for the fresh-water shells whose fossil remains are now found there. Great changes, evidently, have been made in the direction of all of our water courses before they found their present levels."

The valley floor at Layman (Pl. XII, *B*) is not so large as that at Barlow, but it did not carry so large a stream. In several fields on the farms located in this valley the old floor still shows under cultivation a black valley soil, and the author was informed by Mr. J. A. Gage, of Layman, that at one place there is a deep muck, from which much decayed wood has been taken, and the waters issuing therefrom have a very disagreeable odor.

The old floor at Fleming and its associated valley, still smaller than the others, probably carried a smaller stream. The full depths of the silt deposits that cover these remnants of the valley floors were not determined, as all the wells examined were very shallow, reaching a sufficient water supply very near the surface. The bordering hills associated with these old valleys are very low and well graded, and carry very deep soils at the present time, where not exposed to the strong erosion of the present drainage.

WATERTOWN VALLEY.

Not directly in this divide, but associated with the Wolf Creek Basin, is another abandoned valley floor near Watertown. This floor is about 2 miles northeast of the town and about a mile east of the South Fork of Wolf Creek. Rainbow Creek heads on this floor. Whether all or only a part of the stream which occupied this Rainbow Creek Valley drained over this floor is as yet undetermined. Whether or not there were other cols on the Muskingum below Lowell, and whether the reversed Rainbow Creek carried a section of the present Muskingum, will require very careful detailed work to determine, as the erosion of the valley of the Muskingum has been so great in this portion that almost every trace of such cols has been lost. There are some indications in the characters of the divides which would seem to locate one such below the mouth of Bear Run. If this should be subsequently determined it would follow that both Cat Run and Bear Run drained through Rainbow Creek reversed and over the old Watertown Valley floor. The location of this col is not indicated on the map, because it was not considered sufficiently well established.

In the divide separating the waters of Wolf Creek from the Muskingum, immediately south of Roxbury, there is now a very low col, and while it presents few features characteristic of most of the old valley remnants, still it seems quite certain that it represents the location of a very short section of an abandoned valley. The divide at this point is so narrow and the amount of erosion of the large streams on each side is so great (about 150 feet) that nearly all the old valley characters have been lost, but judging from the location of the Meigs Creek col on the Muskingum, a few miles below, it seems evident that the old stream which headed at the Bluerock col must have originally flowed through this low gap.

Along the valley of Wolf Creek and its South Fork are well-preserved remnants of the old gradation plain, which show a marked gradient to the south and. pass directly into the floors of the Layman and Barlow valleys, as indicated on the map (Pl. XI), while the valley at Fleming carried but a small portion of the headwater drainage of the South Fork. The old gradation plain can be traced down the valley of the Little Hocking and its east branch to the plains at the mouth of the stream already referred to in the discussion of the characters of the Ohio Valley.

This gradation plain continued westward into an abandoned valley at Torch, which now forms the divide between two small streams, one flowing westward into the Hocking and the other eastward into the Ohio (Pl. V, *A*). This valley seems to be somewhat narrower than usual for an old valley remnant, but this appearance is probably due to the fact that the silt accumulations are not as thick here as in many other places, and so the valley walls come down closer to the bottom of the broad V. The floor of the valley has been sectioned in numerous places by the Baltimore and Ohio Southwestern Railroad in the construction of its roadbed between the valleys of the Hocking and the Ohio at Little Hocking. The best section exposed is but a few rods west of Torch, where the cut is about 25 feet deep and is located very near the center of the old valley, and in the present crest line. The section above the track shows about 15 feet of very fine clay, scattered through which are small decayed pebbles. Except for the absence of foreign material this clay resembles very much a blue glacial till. No lamination was observed and it was thought to be a very deep residual soil, or possibly a river muck. Above this clay is a layer of from 2 to 3 feet of river gravel, composed mostly of small material varying from a quarter of an inch to 4 inches in diameter, and mostly flattish or lenticular in form. Its local

A.

C.

E.

F.

OLD VALLEY NEAR McARTHUR JUNCTION.—*B.* OLD VALLEY AT LAYMAN.—*C.* OLD VALLEY AT BARLOW.—*D.* OLD VALLEY AT FLATWOODS, EAST OF POMEROY, OHIO.—*E.* ENTRANCE OF THE OLD VALLEY INTO THE OHIO VALLEY NEAR RACINE.—*F.* OLD DUTCH FLATS VALLEY, WEST VIRGINIA.

origin from the Carboniferous sandstones and shales is very evident. The sandstone pebbles are more equiaxial than the pebbles of the shales. All of this gravel is so thoroughly decayed that good-sized pebbles can be easily crushed between the fingers. The section did not show any well-marked evidence of shingling, but was very certainly stream made and stream laid. Above the gravel is about a foot of rather red clay soil and above that some 6 or 7 feet of loess-like silt.

The bed rock is not revealed in the bottom of the cut, so that the exact depth of the filling was not determined; however, the bed rock is thought to be not very far below the railroad track judging from other sections to the east and west which do not show so much clay but which do cut into the rock beneath. In some of these other cuts the gravel lies directly upon the decayed rock surface without the thick clay beneath. One of these sections, about a mile east of Torch, shows about 8 feet of a sandy clay, graduating into the much-decayed underlying rock and overlain by about 2 feet of gravel, and this is overlain by about 4 feet of reddish clay, and above this is some 6 feet of loess-like silt. Immediately below the silt there were observed some evidences of an old soil line. The reddish clay seems to have been deposited over the top of the gravels at a much earlier date than the deposition of the silt above, as would be indicated by this soil line. This suggests that the channel was shifted after the gravels were laid down in the immediate bed of the stream, that the gravels were subsequently buried under flood-plain deposits, and that the reddish clays were part of the bottom land bordering the immediate channel of the stream.

Both east and west of Torch the old valley floor is deeply cut by recent erosion into many very picturesque ravines and gorges. This is especially true on the western sides. The railroad follows one of these ravines from the valley of the Hocking to the old valley floor, making a grade of about 125 feet in 2 miles, while the grade is not so steep from this crest line to Little Hocking. This places the old floor at this point at about 150 feet above the Ohio, or 690 feet above tide.

COOLVILLE AND TUPPER PLAINS VALLEYS.

The old valley extends westward to the Hocking Valley at Coolville, which it crosses at an elevation of about 150 feet above the Hocking River. A section of the deposits on the floor of the valley at Coolville is shown in the main street of the village, where the gravels are revealed to a depth of about 4 feet with about 6 feet of silt above them. In this exposure the gravels are beautifully shingled to the southwest, showing beyond all question the direction of flow of the old stream.

The old valley extending southward from Coolville is a very conspicuous feature of the topography. For a considerable distance between Coolville and Tuppers Plains the floor of the valley is very deeply eroded by a small tributary of the Ohio, but is well preserved at the water parting at Tuppers Plains. The exact elevation of the floor at this point was not determined, as no well sections were found which penetrated the gravels. In the lower portion of the village a number of wells reached the gravel beds at depths of from 20 to 25 feet, passing through fine silts, and, in some cases, through beds of quicksand.

On account of the deep erosion in the basin of Shade River the course of the old valley is more obscured, and yet it can be traced by means of the few benches which are left along the present stream valleys. It crosses from the East Fork of the Shade to the Middle Fork, 2 miles above Chester.

VALLEYS AT HEADWATERS OF FEDERAL CREEK AND HOCKING RIVER.

Two other remnants of old valley floors have been observed in eastern Athens County, one extending between the headwaters of Federal Creek and the Hocking, and the other between the headwaters of the East Fork of Shade River and the Hocking. The valley floor at the head of Federal Creek stands at an elevation of about 730 feet above tide, or 140 feet above the present Hocking, and remnants of its southward extension are found on numerous benches along the Hocking River as far down as Guysville, where they turn to the south and connect with the other valley floor. The other valley is narrow and the table-land rises about 175 feet above the deposits which lie on its floor. One well in this valley is walled for a depth of 40 feet, but no data were obtained with reference to the character of the deposits passed through. A bed of fine gravels containing numerous quartz pebbles was found at an elevation of about 697 feet above tide, which was thought to represent very nearly the elevation of the old valley floor. It is interesting to note the presence of these quartz pebbles in the gravels of this valley floor, for they become an important feature in the gravels of the floors farther southwest. They seem beyond all question to be locally derived from coarse conglomerates which are exposed in this region.

CHESTER AND OHIO FLATWOODS VALLEYS.

From Chester, on Shade River, the old valley can easily be followed southward to the waters of Rays Run (Pl. XIII). About a mile south of the village are many fine sections of the gravels which here rest on the floor of the valley; these are now at an elevation of about 680 feet above tide. Beside the pebbles of the sandstones and shales of the neighboring hills, there are in these gravels numerous quartz pebbles, most of them of about uniform size and about one-half inch in diameter, on an average. Like all the gravels resting on these old floors, these gravels here are very much decayed and the quartz pebbles are coated with a coloring of iron which gives them a very aged and rusty appearance. The valley here swings westward, following the headwaters of Rays Run to the flat plains

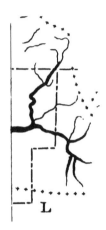

L

B O

DRAINAGE MODIFICATIONS IN MEIGS AND GAL
AND CABELL COUNTIES
B

Elevat

known locally as the Flatwoods (Pl. VI, *D*, and Pl. XII, *D*). At this point the valley is about one-half mile wide and hills rise by gradual slopes up to elevations of 175 feet above the valley floor. The headwaters of Thomas Fork of Leading Creek are on this flat section, and they follow the valley westward nearly to the Meigs County fair grounds, north of Pomeroy, where they leave the valley and cut through the old valley walls in a very deep gorge, while the valley makes a very sharp bend, passing southeastward somewhat parallel to the adjacent valley of the Ohio and only about a mile north of it. This section of the valley is locally known as the Neese settlement. It again bends to the south and merges with the Ohio valley near Racine (Pl. XII, *E*). In the vicinity of the Meigs County fair grounds it is estimated that the old valley floor stands about 160 feet above the Ohio, or 670 feet above tide. It has here suffered very extensive erosion, and undoubtedly the original deposits on the floor of the valley have been very largely removed; however, at a few protected points, sections of the old gravels were observed.

The gravels here also contain a new element in the shape of small nodules of kidney iron ore. These are about the size of beans, and form only a very small percentage of the gravels. The source of this ore is to be found in the thick beds of red clay which occur in the bordering hills at elevations of about 100 feet above the valleys. In some of the ravines east of this locality this hematite ore was concentrated in the bed of the streams during the long period of erosion of the deep valleys, and in the early days was collected for furnace purposes, but the supply was soon exhausted, as there is no large vein from which the material is derived, for it consists only of kidney nodules distributed through the red clay. The abundance of this iron ore has contributed an important and interesting factor to the deposits on the old valley floors farther southwestward, and assists very materially in the correlations of the directions of the streams.

DUTCH FLATS VALLEY, WEST VIRGINIA.

To give an adequate description of the interesting features of this region would require the carefully prepared maps of a topographic survey, and only the briefest outline of general features can be indicated at present. That the old valley crossed the present position of the Ohio at this point is manifest from its southward continuation into West Virginia. On account of the extensive erosion in connection with the valley of the Ohio, and of the location of the oxbow channels, of which there seems to have been several, it is difficult to determine the exact position of the old stream at this point. A marked example of these great oxbows is found near Hartford, where there is an old valley which passes around the river hills and returns to the river again 3 miles below. It is evident that this oxbow has been occupied by the Ohio during its maximum flood-water stages, as there are very extensive deposits of sand in the valley, and these lead across the divide between this valley and the valley of the Ohio below Pomeroy at several low gaps, the most marked of which is about a mile south of Mason City and at an elevation of 160 feet above the river, or 677 feet above tide.

From this Hartford oxbow the old valley extends southward several miles and its floor is deeply cut out by recent drainage lines which now flow westward through narrow gorges across the divide. This old valley is, on an average, about three-fourths of a mile in width, and the bordering hills rise some 175 feet above it to the table-land. The somewhat meandering course of the valley is indicated on the map (Pl. XIII). Its floor is covered with gravels, which furnish an inexhaustible supply of water, as do the gravels of most of the other valley remnants. These gravels are often cemented with iron and lime salts, and are quite like a conglomerate in character.

Besides the sandstone and shale elements, the rusty quartz pebbles which have been mentioned in connection with the old valleys on the north side of the Ohio are here a very marked feature, as is also the numerous nodules of hematite ore. In three instances a very imperfect shingling was observed. Above these gravels there is a deposit of sands and silts, often as much as 40 or 50 feet in depth; the surface portions of the silts are white in color and give the name of white clays to the lands in the vicinity of the upper or Dutch Flats (Pl. VI, E, and Pl. XII, F). Here the surface of the deposits now stands at an elevation of about 685 feet above tide and the old floor is about 15 to 20 feet lower. At Dutch Flats the valley turns westward, and, after making several bends, passes out into the great oxbow in the valley of the Ohio, opposite Cheshire. From the relation of the valley walls, it appears that the old stream passed along the southern side of this oxbow and reached the present position of the Ohio, a short distance above Point Pleasant, opposite the mouth of Campaign Creek. The name Dutch Flats is suggested for this old high-level valley, which resembles in many ways the old Teays Valley.

Throughout the valley the upper slopes are very conspicuously marked by the slender wave terraces, which have already been mentioned in the description of the general topographic features of the lowland section. These present at least three series, extending to an elevation of 850 feet above tide. On a projecting point of the valley wall it was possible to observe these benches through an arc of 260°, and by level observations it was determined that they are practically horizontal at the present time, while the dip of the strata is about 20 feet to the mile, so that it is very evident from these observations, as well as very numerous ones in other districts, that these are not structural terraces due to differential degradation.

About a mile south of Hartford there is a tributary to this old valley which comes in from the eastern side and is now occupied by the waters of a small run.

C.

E.

G

A. OLD VALLEY AT KEYSTONE—B. OLD VALLEY AT PORTER.—C OLD VALLEY AT RODNEY—D. OLD VALLEY AT CENTERVILLE—E A RIDGE ROAD.—F. THE OHIO VALLEY BELOW PORTSMOUTH.—G. VALLEY OF THE LITTLE MUSKINGUM, SHOWING TERTIARY PENEPLAIN AND THREE-WAVE TERRACES ON THE VALLEY WALL.

This valley extends eastward and comes out to the Ohio again opposite Racine. Its floor and the deposits upon it, and its walls, resemble very perfectly those of the larger valley just described, and its direct continuation on the Ohio side of the river is very manifest when observed from the high hills which border the valley. The old gravels were found on this floor at about 680 feet above tide, under about 20 feet of silts.

POINT PLEASANT VALLEY.

At the confluence of the Dutch Flats with the present valley of the Ohio, at the mouth of Campaign Creek, it received another tributary on the southern side, which is now represented by the high-level valley floor east of Point Pleasant, and which represents the old valley of the lower section of the Kanawha River before it crossed the col at Point Pleasant. The old gravels were found under about 20 feet of silt on the old valley floor east of Point Pleasant, at an elevation of about 689 feet above tide, or 177 feet above the Ohio River.

Extending back from the Ohio River opposite Point Pleasant, on the Ohio side, there is another old valley, which has its floor at an elevation of about 670 feet above tide, and which passes northward to the valley of Campaign Creek a little above its mouth. This floor was probably occupied by the lower waters of Campaign Creek when it was flowing at that elevation.

RODNEY VALLEY.

From Point Pleasant the old valley extended a long the line of the present Ohio to the mouth of Chickamauga Creek, just above Gallipolis, where the old gradation plains are well marked at an elevation of about 674 feet above tide. From the general mature features of the Chickamauga Valley it is very evident that it represents the continuation of the line of the old drainage. The remnants of the old gradation plains extend up the valley to the vicinity of Kerr, where they merge into a remnant of an old valley floor which extends to the valley of the Raccoon, past the village of Rodney (Pl. XIV, C). Here the old floor is covered with silts and sands, as shown by well data, to depths of at least 20 to 30 feet. This would make the old floor at this point about 670 feet above tide.

RIO GRANDE AND CENTERVILLE VALLEYS.

From Rodney the valley extends to the west as a well-marked topographic feature, and its old floor is represented by numerous remnants which are preserved along the line of the present Raccoon Creek to the vicinity of Rio Grande, where the old floor stands at 670 feet, or a little less, above tide.

From Rio Grande the old topographic features can be traced westward up a branch of Raccoon Creek to a well-preserved valley floor in the vicinity of Centerville

(Pl. XIV, *D*), where the old valley crosses the present divide between the waters of Raccoon and Grassy Fork of Symmes Creek. Near Rio Grande a section of the old river gravels was observed at the elevation of 670 feet above tide. These were distinctly shingled toward the west, were resting on a decayed rock surface, and overlain with about 10 feet of a sandy yellow silt.

BEDWELL VALLEY.

In the vicinity of Bedwell and Porter, on the headwater tributaries of Campaign Creek, there are extensive remnants of an old gradation plain with a well-preserved old valley floor. A little northeast of Porter, on Campaign Creek, this plain is at an elevation of about 715 feet above tide, and slopes westward under the silts on the old floor to the vicinity of Bedwell, where the surface of the silts is about 698 feet above tide, and where the silts are at least 20 to 25 feet thick. From Bedwell the old valley extends along Barren Creek and joins the present Raccoon Creek near Harrisburg. Along Raccoon Creek from Harrisburg to Rio Grande the old gradation plain is well marked by numerous remnants, although it has been very extensively eroded by the present drainage lines.

CACKLEY SWAMPS VALLEY.

From Centerville the old valley continued northwestward up the valley of Grassy Fork to the open flat lands in the vicinity of the Cackley Swamps, south of Madison Furnace. The valley walls in this region are more completely graded than in the upper sections, and at very much lower angles, and the hills immediately bordering the valley rise about 150 to 200 feet above its floor. The amount of silting in this region has been very great, and there are numerous minor modifications in the streams on both the Grassy Fork and some of its smaller tributaries, but these could be shown only on a very detailed map.

CAMBA, KEYSTONE, AND JACKSON VALLEYS.

From the region of the Cackley Swamps the old valley extends along the line of the Portsmouth division of the Baltimore and Ohio Southwestern Railroad, past Camba station, and on to the headwaters of the Salt Creek Valley at Keystone. The deposits in the valley at Camba are very deep, and it is back of these that the Cackley Swamps are found, which now are the headwaters of Grassy Fork of Symmes Creek. A well, sunk close to the railroad in the cut, penetrated 95 feet of blue clays and sands, and struck rock at that depth. This would indicate the valley floor to be about 650 feet above tide. The surface of the deposits at Camba is about 125 feet above the present surface of the valley at Keystone, the railroad having a grade in this section of about 100 feet to the mile. The valley here is about three-quarters of a mile wide, and its floor is very flat all the way from Keystone to Jackson (Pl. XIV, A). Indeed, it is so flat that it was, in the early days, almost entirely swamp area, many thousands of acres in extent, but most of the swamp land has been reclaimed by cutting the timber and laying drain tiles. At Keystone are located the wells of the waterworks system for the city of Jackson. These wells penetrate from 12 to 20 feet of fine silts and then enter an extensive gravel stratum, which, as far as learned, was not entirely penetrated, so that the exact elevation of the rock floor was not determined. At this point also the old valley receives, on its eastern side, a large tributary of about its own size, entering from the direction of Rocky Hill.

Beyond Keystone the valley is drained by the South Fork of Salt Creek, which enters it about 2 miles past Jackson. At this point the creek turns abruptly to the north and cuts through a very narrow valley extending across the high table-land bordering the old valley, while the old valley continues to the west along one of the smaller tributaries of Salt Creek reversed, and enters the old California Valley in the vicinity of Glade. The features below Keystone and also of the tributary which enters this old valley at Keystone are shown on the map (XVI).

SALINA PLAINS VALLEY.

A few miles above Athens there is a deserted valley known as the Plains, or the Salina Plains, which occupies a position a mile or so back from the present Hocking River. The relations of this old valley are shown more in detail on Pl. XV, in which the dotted lines mark the approximate width of the valleys measured at the contact line between the present valley floors and the valley walls, the continuous black lines indicate the crest lines of the principal ridges, and the dotted areas the position of gravel terraces. Sunday Creek enters the Hocking at Chauncey, and its valley is about three-quarters of a mile in width at this point. Numerous remnants of the old valley floor are preserved in this vicinity, which are now at an elevation of about 725 feet above tide. From the configuration of the valley walls and the surface of the old upland, it is very evident that the old valley of Sunday Creek continued directly across the present valley of the Hocking into the valley of the Plains. Just east of Beaumont a remnant of this old valley floor is still preserved, while east of Chauncey the valley of the Hocking is very narrow and its walls are very precipitous. The interpretations in this region were first suggested by Professor Chapin and were published in the Annual Report of the Ohio State Academy of Science, in which he showed that the Hocking River had cut across this Salina col after the desertion of the Plains valley. The Plains is a narrow belt of country lying south of Beaumont and bending eastward to the Hocking. The middle section of the Plains is very level, and no apparent slope could be determined on the upper surface without careful spirit leveling. It is drained by two small streams, one flowing northward and the other eastward. These streams have cut very deep into the Plains near their lower portions, and show the whole Plains to be an old valley, filled about 100 feet above the present drainage with glacial gravel. Near Beaumont the east wall of the valley presents a vertical cliff 40 or 50 feet in height, which suggests that the old valley floor may have been trenched before the gravels were placed in it; but this may not be true, as it is the only point where such features are shown at all. At most parts of the Plains, bordering hills rise gradually from the surface of the plain and assume the characters of the old slopes of the upland topography.

Directly south of the point where the Plains turn to the east there is a broad depression leading across into the valley of Little Factory Creek. The floor of this depression is about 40 feet above the level of the Plains and about 10 or 15 feet below the rock bench east of Beaumont which marks the level of the old gradation plain at that point.

The valley of the east limb of the Plains is somewhat narrower than the section directly south of Beaumont, which continues across the old valley floor on to the Little Factory. As indicated, the gravels which fill the Plains valley are similar to those which extend down the valley of the Hocking from the morainic tracts near its head-waters. This gravel is extensively used for road purposes, and in these banks some fine sections are exposed. While in general the gravels resemble those of the other terrace deposits characteristic of the through-flowing streams, and no difference in composition was certainly determined, a study of the amount of weathering which these gravels have suffered near the surface indicates that they are weathered to a depth of from 4 to 10 feet, while the gravel terraces in the other valleys which head in the morainic belts of the last invasion are scarcely weathered below a depth of from 2 to 3 feet. In the upper section of this weathered zone the gravels are almost completely decayed, even including the more resistant igneous elements, and the decomposition decreases rather uniformly through the zone, with a well-marked line between the upper weathered portion and the lower undecayed portion. The cementing of the lower portion by the lime and iron salts dissolved from the zone above has not taken place to any large extent, so that the gravels are easily worked by the pick and shovel. In the mouth of Sunday Creek Valley, and extending up that valley for more than a mile, there are remnants of a gravel terrace which show all of the characteristics of the gravels of the Plains region. These are also found in the section of the Hocking Valley below the Salina col and can be traced by smaller remnants down the valley of the Hocking to the great terrace in the great bend of the Hocking Valley opposite Athens, upon which the State asylum is located.

As already indicated, the old valley floor south of the Plains is somewhat above the level of the Plains gravels. The gravels do not extend over it, but upon this floor

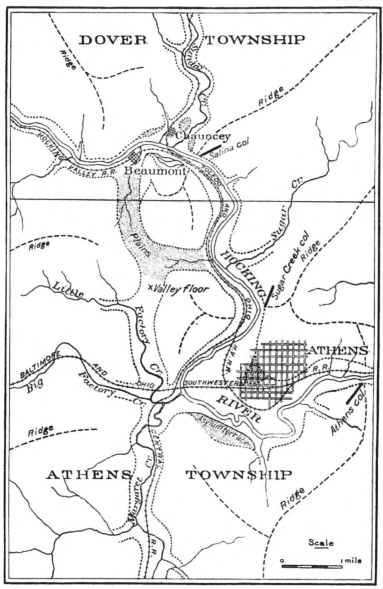

DRAINAGE MAP OF ATHENS AND PART OF DOVER TOWNSHIPS, ATHENS COUNTY, OHIO.

there is a deposit of fine silt, which was seen in section to a thickness of about 12 feet, and under this silt is a bed of gravel about 2 feet thick resting on the decayed rock surface. These gravels resemble almost perfectly those which are characteristic of the old valley floors already described. From the mouth of the Little Factory the old valley is seen to continue directly southward along the line of the reversed Margaret Creek. This is clearly indicated from the general contour of the old upland and the configuration of the bordering valley walls. The present Hocking is now flowing on deposits which fill its former channel to a depth of about 60 feet at Athens, according to Professor Chapin, as shown in numerous well borings on the present flood plain of the stream. The old divide cut through by the present Hocking below Athens, at the Athens col, has already been referred to in connection with the characters of the Hocking Valley.

The history of the region seems to have been as follows: Starting with the old upland plain as a base, the first stage was the development of the matured drainage lines represented by the Sunday Creek Valley above the level of the old valley floor. During this stage Sunday Creek flowed directly southward over the Salina Plains and the valley floor, on into the Little Factory Valley, and thence through Margaret Creek Valley. This was the main axial drainage way of the region. It received a considerable tributary at Beaumont, which headed at the Lick Run col above Nelsonville, and followed the present valley of the Hocking. Sugar Creek probably crossed the present position of the Hocking and entered the old valley in the lower section of the Plains region, while the basin at Athens drained from the Athens col into the old valley at the mouth of Margaret Creek. This condition seems to have been terminated by the deposition of the silts on the valley floor south of the Plains, and also on the old valley floor to be described later in the valley south of this point at Albany, which caused the deflection of the drainage across the Salina, Sugar Creek, and Athens cols. After the modifications thus indicated, the streams were rejuvenated, and new lines were cut to a depth much greater than that of the present streams. Next followed a stage in which the valley of the Hocking was supplied with a load of glacial material, which accumulated along the valley until, in the vicinity of Beaumont, the stream found two ways open to it, one over the old valley floor in the Plains region and the other around the present line of the valley. That the gorging of the material took place here seems to be very clearly shown by the fact that the gravel deposits were built up such a long distance into the mouth of the Sunday Creek Valley, which must have taken place in comparatively quiet water and against whatever current there may have been in the Sunday Creek Valley from the drainage of its basin. Next followed a stage in which the stream was relieved of its burden of gravel and began to clear out its channel. At the initiation of this stage the river chose the route past the Salina col rather than around the Plains Valley, and, during this cycle, cut its channel to its present level, removing almost completely the old filling within its valley, except at favorable points on the inner sides of the curves of its channel, although the Asylum terrace seems to be an exception to this rule. The erosion of the great basin-like portion of the valley at Athens was probably due to the attempt of the river to follow its old line down the Margaret Creek Valley, and in making the sharp bend at this point it cut back the hills on the southern side, leaving the extensive plain on the northern side.

As previously stated, there appear to be no gravel trains in the valley of the Hocking which correspond in characters to the recent terraces in the valleys of the other through-flowing streams, and this is to be noted in this region about Athens.

ALBANY VALLEY.

This old valley, when followed southward up the Margaret Creek Valley (Pl. XVI), is seen to connect with the well-preserved valley in the vicinity of Albany. The upper surface of the deposits in this valley now stands about 775 feet above tide, while numerous wells penetrate the deposits to depths of 30 or 40 feet and often encounter buried wood and vegetable accumulations. Measured across the upper surface of the clays, this old valley is here nearly a mile wide. It bends to the west and passes along a small tributary of Raccoon Creek, where it joins the valley of that stream. The old valley at Albany is now the headwater district for the drainage of Margaret Creek flowing to the Hocking, of Long Run flowing to the Raccoon, and of Leading Creek flowing to the Ohio. The headwaters of Leading Creek pass through a very narrow gorge as they leave the valley along its southern side.

VINTON VALLEY.

Following the topographic features which mark the line of this old Albany Valley, it is seen to pass down the present valley of the Raccoon to its junction with Elk Fork and then turn northwestward up that stream to the vicinity of Vinton station. Along the portion of the valley which is now occupied by the Raccoon there are one or two well-marked gradation plains which lead off from the valley on the eastern side and pass in the direction of the headwater tributaries of Leading Creek, but these were not traced out completely; it is evident, however, that there are numerous modifications to be worked out on this stream.

WILKESVILLE VALLEY.

At the junction of the Raccoon with Elk Fork, this old valley received a tributary from the south which followed along a portion of the present Raccoon, reversed, to the great bend in the present stream below Radcliff. The old gradation plains here

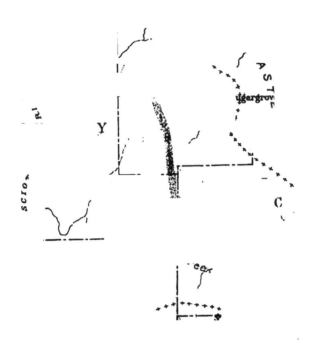

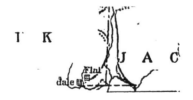

DRAINAGE MODIFICATIONS IN HOCKING V
BY W.
Sc

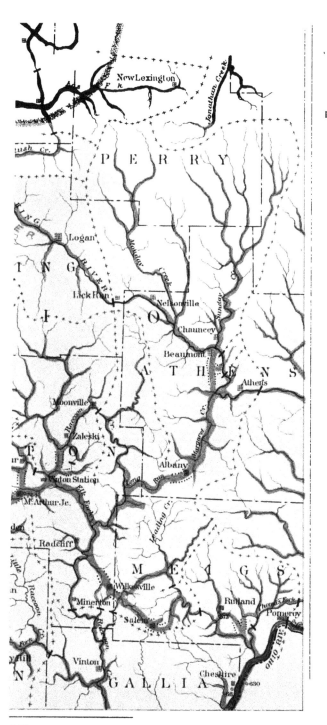

Pre Glacial drainage

Pre-Glacial cols and divides

Pre-Glacial valley floors

Glacial boundary

Observation stations
(direction of view)

Elevation above tide

ON, AND ADJACENT COUNTIES, OHIO
HT

leave the valley of the Raccoon and pass out toward the southeast and over the remnant of an old valley floor to Wilkesville. Just east of the village the old valley is very heavily silted, and the district is now represented by an extensive swamp tract. At this point the valley seems to have divided, one section running off to the northeast and taking the drainage of a section of Leading Creek, the other passing off to the southeast over a well-preserved valley floor at Salem, thence swinging around a curve to the northeast and connecting with another small section of the Leading Creek Valley at that point, along the line of Parkers Fork.

Near Vinton the old valley received a considerable tributary from the north, which carried the waters of the Upper Raccoon Basin above Zaleski and to the Moonville col. This old valley is occupied by the Baltimore and Ohio Southwestern Railroad. In passing over the divide a cut about 15 feet in depth was made in the old silt deposits, the roadbed being here 738 feet above tide. The valley is here about a half mile in width. The extensive silting in this region has effected numerous minor changes of drainage on the smaller streams, notably the modification of Wheelabout Creek, which originally passed over an old valley floor near Prattsville.

M'ARTHUR JUNCTION VALLEY.

From Vinton the valley extends westward to McArthur Junction, and it is here occupied by both the Baltimore and Ohio Southwestern and the Hocking Valley railroads. The latter railroad crosses another deserted valley floor between McArthur and McArthur Junction, with a summit height of 761 feet above tide. Between McArthur Junction and Hamden (Pl. XII, A) the old valley shows its mature characteristics most clearly, and these can easily be observed from the passing train. The valley is here nearly three-fourths of a mile in width, and its floor is heavily silted; the exact elevation of the old rock floor was not determined satisfactorily, as the data obtained were rather conflicting.

WELLSTON VALLEY.

From Hamden the old valley passes around a high portion of the table-land which separates it from the present valley of the Little Raccoon; it then bends southward, and is occupied by the city of Wellston. The floor of the valley is here much cut up by the drainage lines tributary to the Little Raccoon, but it is evident that the sections of these streams between the valley and the Little Raccoon are all reversed at the present time. The headwaters of Middle Fork of Salt Creek have also eaten into the old divide and caused the watershed to migrate toward the old valley to such an extent that the position of the old divide is represented by a very narrow gorge, through which the drainage passes down Salt Creek.

From Wellston the valley runs almost due south to the vicinity of Berlin, where the floor of the valley is a water parting for another tributary of the Little Raccoon, Dickson Fork. At Berlin the valley is fully three-fourths of a mile wide, measured on the surface of the silt fillings, which are here rather thick, although their exact depth was not determined. Much of this area is now represented by swamp lands which have been reclaimed by draining.

From the Berlin Flats the valley continues southward and is drained by Dickson Fork of the Little Raccoon. The old gradation plains along this section of the valley are well preserved, and where Dickson Fork leaves the valley they stand at an elevation of about 50 feet above the stream. There are also well-marked remnants of the old gradation plain on Dickson Fork, and these rise rapidly passing down the creek, showing that the old drainage was exactly the reverse of the present. From this point the valley passes somewhat to the southwest, beginning a great curve which it makes to join the valley already described at Keystone. At Rocky Hill, which is the summit of the Cincinnati, Hamilton and Dayton Railway, the old valley floor now stands at an elevation of about 700 feet above tide.

RECONSTRUCTION OF THE OLD DRAINAGE SYSTEM.

Having discussed briefly the character and distribution of the old high-level valley floors, with the remnants of the gradation plains along many of the present stream valleys, and the location of the old cols crossed by the present streams in their development, it is possible to reconstruct with considerable accuracy the drainage system of the cycle of erosion which is represented by the old valleys.

As at present, the Kanawha River was the principal axis of drainage of this basin. Its old course from St. Albans across Teays Valley to the Ohio and thence through the Flatwoods Valley and northward along the present Ohio to Wheelersburg, and from this point through the old California Valley to the Scioto at Waverly, seems to be established beyond all question. This restoration was suggested by Mr. Frank Leverett and the author,[a] while Teays Valley was long ago recognized by numerous other observers as an abandoned portion of the Kanawha. Mr. Leverett, in the paper cited, suggests that from the Scioto at Waverly the old course of this stream may possibly have been down the Scioto to the Ohio and thence directly down the Ohio; yet on his map,[b] he has located an old col on the Ohio above Vanceburg, which, of course, is inconsistent with the view expressed. It seems certain to the writer that this col is located farther down the Ohio (a mile or so above Manchester), as indicated on the map (Pl. XVII). It is therefore evident that at Waverly this old

Kanawha drainage received a large tributary from the south, which included the waters of Salt Creek and Kinniconick Creek, of Lewis County, Ky., with the minor tributaries, and that the old direction of this drainage was along the reversed Ohio to Portsmouth, and from Portsmouth along the reversed Scioto to Waverly, as originally described by the writer in his earlier publication.

With this restoration of the old drainage along the lower course of the Scioto, it is to be noted that all the tributary drainage falls into normal relations, which fact, while not conclusive proof, yet adds weight to the correctness of the interpretation.

As the present drainage of the Kanawha and the section of the Ohio so nearly corresponds with the old drainage line through the southern division of the basin, it is to be expected that most of the streams tributary to this line would still exhibit their normal relations; that they do is shown by the tributaries on the south side of the axis, viz., Tygarts Creek, Little Sandy River, Big Sandy River, Twelvepole Creek, Guyandot River, Mud River, Coal River, etc. The Guyandot and Mud rivers were tributary to the old drainage line at Barboursville and Milton, respectively. On the north side of this axis, in the plateau region, the Pocatalico River, the Elk River, and the Gauley River are the principal northern tributaries. Below the mouth of the Pocatalico, which entered the old drainage below St. Albans, all of the lateral drainage lines, while quite normal to the axial line in their arrangement, are very short. They include the small stream which headed at the Crown City col and flowed along the present position of the Ohio to Guyandot, Indian Guyan Creek, the lower section of Symmes Creek, Ice Creek, and Pine Creek, with possibly one or two other smaller tributaries in eastern Scioto County not yet determined, these being tributary to the California Valley section.

The restoration in the northern section of the basin, in its relation to the present drainage, is much more complicated. The headwaters of the axis of this portion of the basin include the Middle Island Creek and Little Muskingum drainage and a small creek which headed at the New Martinsville col and flowed along the present position of the Ohio to Newport, where it joined Middle Island Creek. These streams followed along the present Ohio to Marietta, where they received the lower waters of Duck Creek and the lower waters of the present Muskingum below the Lowell col. From Marietta the old valley followed very closely the present Ohio to Little Hocking, receiving at Parkersburg the drainage corresponding to the present Little Kanawha Basin. At Little Hocking it received a large tributary on the northern side, which consisted of two large branch streams. The eastern one, heading in Morgan and Noble counties, included the Meigs Creek, Olive Creek, and Big Run basins, and all of that section of the Muskingum between the Meigs Creek and Lowell cols. These waters flowed along the South Fork of Wolf Creek reversed, and passed through the old Barlow valley to the present east branch of the Little Hocking. The other tributary drained the western portion of Morgan County, and consisted also of two large branches, one being the stream which headed at the Bluerock col and flowed along the present course of the Muskingum to Roxbury, where it crossed the old Roxbury Valley floor to the waters of Wolf Creek and thence flowed up the valley of that stream to its southernmost bend, where it was joined by the drainage represented by the present upper waters of Wolf Creek. Together these streams crossed the old valley floor at Layman to the headwaters of the Little Hocking, and passed down that stream to its junction with the eastern branch. From Little Hocking the axial drainage passed through the old Torch valley, crossed the present Hocking at Coolville, thence flowed southward past Tuppers Plains to the East Fork of Shade River. Here it received a large tributary on the northern side, which included the Federal Creek Basin and all of that section of the Hocking between the Athens and the Federal Creek cols. From the East Fork of Shade River the axial stream passed across to the Middle Fork and thence down that stream to Chester; thence it followed along the line of the present Rays Run reversed to the Flatwoods valley, east of Pomeroy, where it made a great sigmoid bend past the Meigs County fair grounds, through the Neese settlement, to the Ohio River near Syracuse. Here for a very short distance it conformed to the valley of the Ohio, but left that valley at Hartford and passed southward into West Virginia along the old Dutch Flats Valley. Just below Hartford it received a considerable tributary, composed of the drainage of the Big Sandy Creek and Mill Creek basins, with a possible addition of the drainage of the present Ohio up to the Hockingport col; but the evidence in hand seems to indicate, although it is not conclusive, that this section of the present Ohio was divided into several smaller basins which drained directly eastward into the old valley. These are indicated on the maps.

From the Dutch Flats Valley the old drainage continued westward to the present Ohio at the mouth of Campaign Creek. The exact relation of its northern tributaries at this point is somewhat doubtful, as the drainage of the section of the Ohio below Pomeroy and the lower portion of Leading Creek may have entered the old axial line in the present position of the lower Sand Flats, or, possibly, may have followed along the present Ohio to the mouth of the present Campaign Creek. These doubtful relations are also indicated on the map (Pl. XIII). At this point the old stream received a large tributary on its southern side, which drained the basin represented by the lower portion of the Kanawha River below the Winfield col and the Hurricane and Little Hurricane Creek cols. This stream will be called the Point Pleasant River. From this point the old stream passed down the Ohio to the mouth of Chickamauga Creek and up that valley and over the old valley floor at Rodney; thence directly westward to the Raccoon, at Rio Grande. At this point it

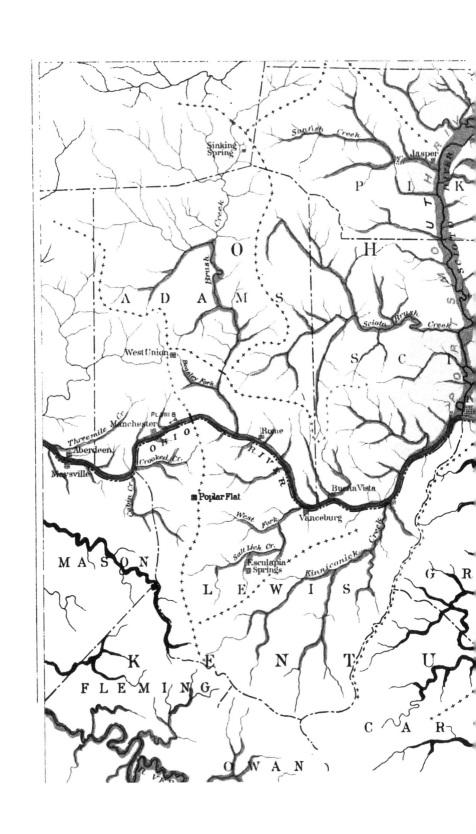

Raccoon Creek

A L

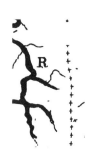

R

. JACKSON, AND LAWRENCE COUNTIES, OHIO,
ES. KENTUCKY, AND VICINITY
3HT

JULIUS BIEN & CO LITH. N

10   15   20 miles

received a tributary on its northern side, including a portion of the present Raccoon Creek below the Minerton col, with its smaller tributaries, and the old Porter Creek, which included the upper waters of the present Campaign Creek. On the southern side it received a much larger tributary—a stream which headed in Little Guyandot Creek and which included the drainage of the Ohio above the Crown City col to the mouth of the Raccoon—and from there this tributary followed up the present Raccoon, receiving a small tributary at Lathrop and another near Patriot, which included the drainage of Symmes Creek between the Marion col and the Evans Mill col.

From Rio Grande the old axial valley passed westward over the Centerville Valley floor into the Grassy Fork Basin, and thence over the old Camba Valley to the great Jackson Valley at Keystone. Between Centerville and Keystone it received several small tributaries made up of sections of the upper waters of the Symmes Creek Basin.

At Keystone the old axial stream received its largest tributary on the northern side. This old river had its headwaters in Monday and Sunday creeks, including that section of the basin of the Hocking between the Lick Run col and the Athens col. It passed from the mouth of Sunday Creek across the Plains Valley, up the Margaret Creek Valley to Albany; thence westward to the valley of the Raccoon, where it received a tributary including the present Raccoon drainage below the Moonville col. It then passed down the present Raccoon to the mouth of the Elk Fork, where it received a large tributary on the southern side, composed of a large portion of the present basin of Leading Creek and of the Raccoon north of the Minerton col. From the mouth of Elk Fork it passed up this stream to the vicinity of Vinton station, where it received, on its northern side, the drainage now represented by the upper waters of the Raccoon above the Moonville col. From this point it passed westward through McArthur Junction, where it received a small tributary including the upper waters of the Elk Fork Basin. From McArthur Junction this river passed southwestward past Hamden, Wellston, Berlin, and Rocky Hill to its confluence with the axial stream. Between McArthur Junction and this point it had many minor tributaries.

From Keystone the old river passed through the Jackson Valley to the old California Valley at Glade (Pl. XVII).

Thus the drainage of the entire basin was collected into a great river at near Waverly in the present Scioto Valley and followed this valley northward to beyond Chillicothe. North of this point the location of the old channel has not been accurately determined on account of the great depth of the drift covering this section of the State. It is probable, however, that it continued northward and joined the old Newark River somewhere within the middle section of the Scioto Basin, as did

Adelphi Creek and the Lancaster Creek, which represent the smaller basins in the angle between the two larger streams.

For convenience of reference it seems best to give names to the larger portions of this drainage, and so the name Chillicothe River is suggested for the major stream, as the city of Chillicothe lies near the mouth of this river as known at the present time. For the main axial stream, including the Kanawha River system, the name Teays River is suggested, as the Teays Valley is the largest remnant of its old course; and for the northern tributary the name Albany River is suggested for the northern branch and Marietta River for the southern branch, as the cities of Albany and Marietta lie near the headwaters of these streams.

The general relations of the old drainage of the entire basin and its relations to the restorations in the neighboring regions are shown on the map (Pl. I.)

The old drainage of this basin presents some interesting peculiarities, and the interpretation of these will enable us to determine some of the deformations of the basin during the development of this old drainage cycle. It is to be noted, however, that the old drainage presents many normal features which are characteristic of streams consequent upon the slope of the plateau on which they are developed, and that the normal characters of this system within the basin are very much more marked than the distribution of the present drainage. While this evidence is not conclusive it certainly is very suggestive of the general truthfulness of the restoration.

The great development of the tributaries of the Marietta River, between Parkersburg and Keystone on the southern side of the stream, as compared to those on the northern side, and a similar development of the streams tributary to the Kanawha between Charleston and Ironton on its southern side, as compared to those on the northern side, suggest the advantage which these tributaries must have had over their complementary streams on opposite sides of the valley. It would seem natural to conclude that the Kanawha was deflected from the old Teays Valley by the piracy of the old stream which it now follows below St. Albans. This suggestion has been proposed by several authors, but the fact seems to be lost sight of that while this divide, the Crown City-Hurricane-Winfield divide, has undoubtedly migrated southward toward the Kanawha, the Ohio in its development has crossed this divide from the north toward the south, while the Kanawha has crossed the divide from the south toward the north. This latter consideration seems to be entirely opposed to the idea that the Kanawha was deflected from Teays Valley by stream piracy, while at the same time we can see, in the relations of the streams to this divide, that its cols were probably very low and hence the deflections of the streams during the stages of modification over these cols in either direction were readily accomplished.

The very crooked courses of the Albany and Marietta rivers suggest that the positions of their valleys were determined by the extensive meanderings of their ancestral streams over the old base-leveled plain, and that this portion of the basin, which now constitutes the lowland section, was more completely base-leveled than the plateau section which was drained by the old Teays River.

RECONSTRUCTION OF THE OLD TOPOGRAPHIC FEATURES.

It seems possible, with the data in hand, to reconstruct the topographic characters of this region before the drainage modifications took place. This may be accomplished by considering the effects of the restoration of the deposits eroded since the period of rejuvenation, of the removal of the deposits of silts along the lines of the old drainage, and of returning the drainage to its old lines. The conditions which would be presented have already been briefly mentioned in connection with the discussion of general topographic features, but it seems well to consider them somewhat more in detail, as an appreciation of these conditions is essential to a proper understanding of the causes and method of modification, from the older to the newer system.

We see in the broad table-lands the form of the general country surface. This consisted of a gently rolling but somewhat hilly topography, and the drainage lines were cut only from 100 to 200 feet below the general surface of the country. The valley floors and slopes were well graded, and the headwater streams had eaten back into the old divides until there were very numerous low cols separating the various minor basins of the region.

It is apparent that under such conditions the obstruction of any of the streams would result in the deflection of the portions of the streams above the obstruction, over the headwater cols, even though the obstruction were not of very great elevation. While it is evident that the region was near the base-level of erosion toward the close of this old cycle, it does not of necessity follow that it was much lower in absolute altitude than at present, although this is strongly suggested, for it may be found, upon the more complete restoration of the lower waters of these old systems, that they had a much longer course to the sea than does the present drainage, in which case the base-level of erosion would stand much higher. It seems to the writer that this should be borne in mind when the base-level condition of the old drainage, which now stands at a higher level than present drainage, is assumed as evidence of subsequent elevation.

RELATION OF PRESENT AND RESTORED DRAINAGE SYSTEMS TO GEOLOGICAL STRUCTURE.

The discussion of this subject is confined to the plateau and lowland areas of the basin, as the relations of the drainage to the geological structures in the Appalachian valley and mountain section of the basin have no direct bearing upon the problem in hand.

As the distribution of the old drainage lines was not heretofore determined, it is evident that all references to the relation of the drainage to the rock structures were based on the assumption that the present drainage lines are of very great age; in fact, in all of the writings up to within a very few years it is assumed without question that the Ohio and its principal tributaries date back to the beginning of the Appalachian elevation, and hence had their origin somewhere in the Permian, as rocks of this age are not represented in this region. From an apparent relation which seems to exist between the position of some of the present river valleys in Ohio to the strike of the strata, it was early suggested by Professor Newberry, and later by Dr. Orton, that these lines of drainage were determined by the relation of the underlying hard and soft beds. Perhaps the most noted example of this coincidence is the Scioto River, and the suggestion has been offered that the great basin of the upper portion of the Scioto has been eroded out of the soft beds of the Ohio shales. It is to be noted, however, that in passing southward the Scioto leaves the region of the Ohio shales a short distance below Chillicothe and passes directly into the face of the escarpment of the Waverly sandstones and conglomerates, and, after making a considerable detour to the east, again returns to the region of the shales near Waverly; but here again it does not follow along the western edge of the outcrop of the Waverly series, but again enters the area of the Waverly and follows that to its mouth at Portsmouth. While it is true that north of Chillicothe the surface of the Ohio shales lies very low and that a very broad basin has been excavated from them, it is equally true that where these shales approach the Ohio River they form there the tops of some of the very highest hills of this section.

Again it is to be noted, in the case of the Muskingum, that the western tributaries of the Tuscarawas, viz, Chippewa Creek, Shade River, the Killbuck, the Mohican, and the Walhonding, all have their headwaters within the area of the Waverly, and cross the western outcrop of the Carboniferous sandstones and conglomerates to the basin of the Tuscarawas. The same is also true of Licking River, which is tributary to the Muskingum at Zanesville.

Again, in the case of the Hocking, it is to be noted that its headwaters lie in the area of the Waverly and that it crosses the outcropping edges of the Carboniferous sandstones and conglomerates on its way to the Ohio, while in the case of the Little

Miami and the Great Miami rivers their courses lie almost parallel to the axis of the eastern limb of the Cincinnati arch, and their direction of flow is against the dip of the axis of this fold.

With reference to the drainage in the northern and western sections of Ohio, it was early recognized that the distribution of the drainage in those sections was determined almost entirely by the surface characters of the drift.

It seems certain, therefore, that the attempt to correlate even the present distribution of the drainage with the underlying structure fails to reveal any direct relationship.

In various passages scattered through the Ohio surveys, from the early investigations of Whittlesey, Andrews, and Locke, and, later, those of Newberry and Orton, reference is made to the difference in the topographic features of portions of the same river valley, and the suggestion is offered that where the streams pass through the areas of the soft shales and less resistant beds of sandstone the valleys are broad and the topographic features those of a mature system, and that where the streams cut through the harder and more resistant beds they are in general in narrow gorges, and numerous instances are referred to in support of this conclusion; but it seems as though the statements were made more to explain the difference in the topographic features of the valley than from an interpretation of the observed facts, for it is evident that in every case of this kind which has been investigated by the author it has been found that the broad and open portions of the valleys are in exactly the same systems of rocks as are the narrow and gorge-like portions.

Perhaps the most common incident referred to in this connection is the cutting of the streams through the resistant rocks of the Logan conglomerate of the upper Waverly series. An example of this is presented in the gorge of the Licking at Black Hand, but within a mile of the Licking gorge and almost directly north of it the old deserted Hanover Valley crosses the outcrop of the Logan conglomerate. The valley is here nearly a mile wide, and the topographic features of the old mature form are well shown. It seems that the idea of the great difference in the age of the wide and rolling portions of the valleys and of the narrow, gorge-like sections had not occurred to these earlier investigators. This appears perfectly natural when we consider that they believed in the general continuity of the course of the streams from a very remote period.

As the general constrictions of the valleys and the precipitous gorge-like characters of these narrow portions have been employed as an important factor in the location of the old cols in the old divides which were cut through by the present streams in their development, and as it is a well-recognized principle, which the older geologists used in their interpretations, that the variation in the hardness of the rocks plays an important part in determining the characters of the valleys which cross alternately the hard and soft beds, it is of importance to note just what relation these gorge-like sections of the present river valleys bear to the outcrop of the hard beds over which they pass.

It is to be noted, in the first place, that with the exception of a very small area on the extreme western side of this basin the plateau region and the lowland region both lie entirely within the area of the Carboniferous beds. It is evident, therefore, that the degree of constriction of these valleys, as explained on the basis of their relation to the hard beds, will be limited by the differences in hardness between the various elements of the Carboniferous series. An examination of these elements does not reveal the fact that there are any large portions of them which stand in striking contrast to any other large portions in their hardness or resistance to weathering and degradation. While it is true that there are beds of harder sandstones and conglomerates interspersed with soft shales and coal, all of these beds are so nearly horizontal in position, and the layers of different hardness are so thin in comparison with the entire thickness of the series, that it would seem difficult to see how the outcropping portion of the thin soft bed would give an opportunity for the extensive erosion of long sections of a valley and the contemporaneous limited gorging of a short section. It would seem as though such a condition would be produced by the erosion of monoclinal beds in which the soft beds were extremely thick and the hard resistant beds were comparatively thin zones.

In the lowland section, where the variations in the characters of the various sections of the same stream course are most abundant, the rocks dip in a general way southeastward. If, therefore, the valley constrictions are due to the crossing of the outcrop of the hard layers of the series by the streams, it should be possible to determine beforehand the characters of the stream valleys along the lines of these outcrops—that is, along the strike; but in no case has this been found possible, and a glance at the maps will indicate, from the great irregularity of the distribution of the eroded cols, which, of course, mark the points of constriction of the present valleys, that there is no relation whatever between these locations and the outcrop of the hard beds. Examination of the rocks at the constricted parts of the valleys shows that they are soft shales about as often as they are hard sandstones. The same thing is true with reference to the distribution of the old valley floors, which show the topographic features characteristic of the broader portions of the present river valleys. It seems certain, therefore, that the variations in the hardness of the different layers of the Carboniferous rocks, and also of the other formations of the region, have had but a very minor influence in determining variations in the topographic features.

It is true that portions of the same valley of the older system often present very great differences in the widths of the valley and in the amount of gradation that has taken place along the valley walls, but in every case investigated it has seemed that the factor which has been most operative in producing these variations has been that of the distribution and spacing of the tributary drainage.

An examination of the relations between the restored drainage of the old cycle and the underlying structure fails to reveal any influence which the latter may have had upon the course of the streams. It may be said that, in general, the basin was drained in the direction opposite to the dip of the strata, and that in this region the drainage from the area of the Carboniferous is across its outcropping edge onto the area of the Waverly, and thence across the outcropping edge of this formation into the region of the Ohio shales.

It seems certain that the courses of the old streams were determined entirely by the slope of the original plain of elevation and the deformations which it has suffered since the inauguration of the original drainage. It may be true that in some minor cases the underlying structure has affected the course of particular streams, but thus far no important examples of such have been reported. It follows, therefore, that if the old streams, in their directions and distribution, show no direct relations to the underlying structure, and that no relation can be determined in connection with the present streams, the latter can in no case be considered as derived from the former through the ordinary processes of stream action and modification depending upon the discovery of underlying structures in the downcutting, which would convert the old consequent streams into subsequent ones. The varying resistance of beds to erosion, as a factor in producing the modifications, is of no value in the present case.

RELATIONS OF OLD AND PRESENT RIVER SYSTEMS.

These relations are expressed in two ways, in a horizontal and in a vertical direction. The horizontal relations are marked by very great irregularities. There seems to be no systematic order which can be traced in the relations between the two cycles. The deflections appear to be as much in one direction as another, so that any suggestion that the deflections have been produced by deformations of the basin seem to be entirely without support in the observed facts. This great irregularity in the directions of modifications is very manifest from even a casual examination of the maps, which show the two systems superposed. This is equally true from field observations. For example, standing on Horizon Hill, a few miles northwest of Marietta, one gets a very complete view of the topographic features for many miles around, and the distribution of the drainage which these features indicate contrasts very strongly with that indicated by a map which shows the actual drainage conditions. The same is true in many other parts of the basin. The general normal characteristics of the old drainage, in contrast with the apparent abnormalities of the present, furnish about the only guide in the field investigations, and it has been necessary to follow very carefully the old topographic features in order to determine the position and direction of the modifications. It is a noteworthy fact that in most cases the new or present lines are more tortuous than the old lines, e. g., the northern deflection of the Raccoon at Zalaski across the Moonville col, while its old line lay directly to the south, and the deflections of the Muskingum across both the Meigs Creek and Lowell cols; also the deflection of the Kanawha from its direct route through Teays Valley to its more distant route to the Ohio at Mount Pleasant. In this connection it is to be noted also that in a few cases the new drainage shows a tendency to cut off some of the longer detours and the angles between the old lines, e. g., the short course of the Ohio from Wheelersburg to the mouth of the Scioto at Portsmouth, for it would appear that the lowest route would have been through the California Valley to Waverly and down the Scioto; and again, the tendency of the new drainage to cut off the short angles of the old is shown in the deflection of Rush Creek at Sugargrove, where it enters the present Hocking, of the Kanawha at Point Pleasant, and of the Little Kanawha at Parkersburg.

Another marked contrast is shown in the relation between the size of the valleys and the size of the streams which now occupy them. In the case of many of the old deserted valleys, the present streams are small and insignificant brooklets, and the effect has been to greatly erode the floor of the old valley. This is shown especially in the old Dutch Flats Valley between Hartford, W. Va., and the Ohio at Cheshire, while the opposite relation is expressed where the old valleys, which were occupied by small streams, are now occupied by very large streams of the present system. In this case the effect has been to almost entirely obliterate the characters of the smaller valleys by the great erosion which has been produced by the present larger streams, and often the larger streams seem to have crossed a number of minor cols within a very short distance, indicating that the position of the larger stream was determined by its wanderings in and out among the smaller headwater streams of the old valleys. This tendency is noticed in the grouping of the cols in the vicinity of Pomeroy and Gallipolis.

The vertical relations are as varied as are the horizontal ones. The present streams are very generally graded, although not perfectly so. The Ohio has a fall of from 592 feet above tide at New Martinsville to 451 feet above tide at Manchester; the Muskingum has a fall of from 700 feet above tide at Zanesville to 570 feet above tide at Marietta; the Hocking has a fall of from 625 feet above tide at Athens to 560 feet above tide at its mouth; Raccoon Creek has a fall from 700 feet above tide at Zalaski to 510 feet above tide at the Ohio; the Scioto has a fall of from 600 feet above tide at Chillicothe to 468 feet above tide at Portsmouth; while the Kanawha has a fall of from 570 feet above tide at Charleston to 512 feet above tide at Point Pleasant. The average grade of all of these larger streams is about 1.7 feet per mile.

The old axial stream of the Teays River basin had a fall of from about 675 feet above tide at St. Albans to about 600 feet above tide at Waverly.   The old Kinniconick Creek, along the tributary which the Ohio followed up in crossing the Manchester col, had a fall of from about 800 feet above tide at Manchester to 600 feet above tide at Waverly.   The old Marietta River had a fall of from about 770 feet above tide at St. Marys to 600 feet above tide at Waverly, and the old Albany River had a fall of from about 725 feet above tide at Chauncey to 600 feet above tide at Waverly.   The grades of these old valleys are given on the assumption that the movement to which the region has been subjected has been of such a character that the grades have remained practically parallel to themselves.   These figures would, therefore, have to be modified to the amount of any deformations which it may be shown that the basin has suffered since the old grades were established, but as they now stand these grades indicate an average fall of about 1.6 feet to the mile, which represents the grade of the lower portions of the old streams, while the grades in the headwater sections were undoubtedly much higher, as the valley floors of the smaller headwater streams are now found at elevations of 250 to 300 feet above the present drainage.   It thus appears that the vertical limits extend through a range of about 300 feet, and these are most marked where the valley floors of the smaller streams of the old system come into a close position with the present larger streams.   The two systems are about equally graded as they now stand, and, from the great diversity in the horizontal distribution, this fact would point toward the comparative stability of the basin during the erosion of both cycles, and certainly is opposed to any marked differential movements within the basin.

In contrast with the extensive cutting of the present large streams below the small old ones are the conditions where the old valley floors are buried under deposits of silts, and the present smaller streams are often as much as 50 to 60 feet above the floors of the old streams.   Examples of this relation occur at the old floors at Barlow and Albany, and at other places.   Between these two extremes there is a perfect series of combinations, which result in a great variety of topographic features.

In this connection it may be noted that in the regions directly north of this basin another combination is even more common than those mentioned—that is, where the old valley floors lie very much below the position of the present drainage, and the old valleys are filled with the drift deposits and silts, often to a depth of many hundreds of feet—but examples of this type are not known to the author within the basin under consideration, as the valley floor of the old Chillicothe River does not approach the present drainage until it passes beyond the limits of the region under discussion.

RELATION OF PRESENT STREAMS TO THEIR VALLEY FLOORS AND TO DEPOSITS WITHIN THE VALLEYS.

As has been indicated in a previous section, the present streams are now running in most of their courses above the rock floors of their present valleys. These floors are buried under extensive deposits of gravels, sands, and silts, which are quite variable in character in the different portions of the basin and also in different portions of the same valley. It is natural to expect that there would be, as there is, a striking difference in the deposits of the through-flowing streams and of the indigenous streams.

In the valleys of the through-flowing streams the lower portions of the deposits consist of alternating beds of gravel, fine sand, and often quicksand, with beds of silts, while the upper portions are composed very largely of well-sorted gravels, with occasional layers of fine silts. A marked example of this is reported by Prof. E. B. Andrews[a] in the structure of the terrace at the mouth of the Muskingum at Marietta. The materials which compose these gravel deposits are of foreign or of glacial origin, and their source can be traced by following up the gravel trains into the morainic belts of the glaciated area. The upper portions of these deposits rise to an elevation of 60 to 80 feet, and in a few cases to even more than 100 feet above the present levels of the streams. The channel ways cut into these deposits are very generally marked by numerous terraces, which represent stages in the process of cutting. This is shown from the observed sections, which exposed the contact zones between the highest terrace plains and those below it, where the structure appears continuous, while if the lower terraces were subsequent deposits in a channel way cut from the upper terraces there should appear an unconformity between the two deposits.

Over the present flood plains of the streams are extensive deposits of silts; between these silts and the terraces there are very generally present marked unconformities. The present streams are running in channel ways cut through the silt deposits. In their meanderings they often swing over against the terraces and even to the rock walls of the valleys. Thus various combinations may occur at particular points between the rock walls, the highest terraces of gravel, the intermediate terraces, and the silt deposits. In the cutting out of the lower terraces from below the level of the highest terraces, the stream has often occupied two channel ways, and in the process of cutting one has gained all the volume of the stream, thus leaving a deserted channel way surrounding an island-like mass of the higher terrace isolated and standing out in the middle of the valley, as in the case already referred to at Cheshire, where there is the old channel west of the village along the bluffs which stand back more than a mile from the river, which courses along the east side of the valley.

In the cutting out of the present channels it has often happened, also, that the meanderings of the stream have thrown the channel so near the old valley wall that the lower rock slopes which were buried by the gravel deposits have been discovered and the present streams are running on a rock floor which may extend only part way or even entirely across the present channel, the old deeply filled valley lying to one side of the present rock floor.   Cases of this kind are very numerous.

During the stage of maximum filling it frequently happened that the deposits reached elevations sufficient to surround greater or less masses of the hills which border the valley, and the stream, when it began the removal of the deposits, occupied the route surrounding these hills.   Soon it discovered the rock in the low col, and in the subsequent down cutting there was developed a narrow rock gorge, and the old valley lies filled with deposits to one side of the mass of hills.

Again, the stream continued to occupy the old channel, and thus the upper terrace deposits are seen to pass around and behind the rock hills, forming the immediate valley walls.   Numerous instances of all of these cases have been observed and described by many authors, and it is needless to repeat them here.

The interpretations of the drainage modifications have not previously gone beyond this point.   From these observations it seems that the history of the through-flowing streams within the present valleys has been quite varied.   The first stage was the extensive erosion of the deep valley floors to a depth of from 50 to 100 feet below the present drainage level.   The length of time involved in this stage would depend somewhat upon the volume and velocity of the stream. In the narrow portions of the valleys, where the terraces are entirely absent or but poorly preserved, the present streams often wash the talus slopes of both sides of the valley, and these slopes lie at the bases of the vertical cliffs and are often as steep as the material which composes them will stand.   From their form and relation to the present streams they suggest that the stream which eroded the deep valley was much more powerful than the present one.   The present river is able to do little or no work in the way of removing the talus, while the old stream, in order to have cut the gorge 50 to 100 feet deeper, must have been superior in its action to the supply of the talus.   It is recognized that the reduction of the grade of a stream of the same volume as the present would produce the same results, but there are other minor features, such as the character of the curves of the valley walls at these narrows, which suggest very strongly, although they do not demonstrate, that the valleys were cut by torrential waters during some stage of their history.   That the rate of cutting was very rapid as compared to the rate of degradation on the sides of the valleys is very certainly shown in the bold cliffs which face the streams, and these are often composed largely of soft shales, in comparison with the gentle and well-graded slopes of the same hills on their faces which look away from the valleys. This truncation of the older features by the newer streams has already been mentioned.

The second stage began with the deposition of the gravels, sands, and silts on the floors of the deep valleys, and continued until they had accumulated to the level of the highest terraces. That there were many minor changes of the conditions during this stage is manifest from the nature of the deposits. These indicate stages of very vigorous drainage when the streams were overloaded with materials and great thicknesses of very coarse gravels were laid down. In the less direct lines of current action were deposited the sands, and by the shifting of the currents these deposits were made to overlie each other in a very irregular way. At times there seems to have been a slack-water condition, when the silt deposits reached many feet in thickness, and these were again followed by the more vigorous drainage and the continued deposition of the coarse gravels. It may be, however, that the deep deposits of silt only represent the dead-water conditions produced by the conflicting currents which carried the coarser gravels, and that they are not indicative of a slack drainage throughout the entire lengths of the valleys. The data in hand are not sufficient to read with certainty the various changes of this stage. It continued to the time of the deposition of the highest terraces. That the gravels of these terraces were deposited in rapidly flowing streams of large volume is very evident. The upper surface of the highest terrace in the Ohio Valley does not show a uniform grade, but rises much higher above the present drainage along some portions of the valley, especially below the mouths of the Muskingum, Hocking, and Scioto, indicating that the supply of material at these points was so great that the current down the Ohio was not able to reduce these local accumulations to a uniform grade.

The third stage was inaugurated by the reduction of the materials supplied to the streams, with a possible increase of their gradients; but it seems sufficient to assume that if the supply of material should cease the streams would at once become eroding instead of depositing streams. From the old deserted channel ways and many minor features connected with the forms and distribution of the intermediate terraces it seems certain that the streams which cut them out of the high-level deposits were very much larger than the present streams. This third stage of erosion is considered to have extended to the time when the volume of water was reduced to near that of the present through-flowing streams. The length of this stage is then represented by the erosion of the deposits from the highest level terrace to near that of the present streams.

The fourth stage includes the activity of the present rivers since the close of the third stage. In this fourth stage the present streams have been depositing silts and building extensive flood plains in some sections of their valleys, while in others they have evidently been spending their energies 'in removing barriers, usually in the form of rock ledges, which obstruct their channels.

A comparison of the work accomplished in each of these four stages, with a view to ascertaining the time relations of each, while attended with many uncertainties, seems to indicate in a very general way that if the fourth stage is considered as 1 the third stage would be represented by about 20, the second stage by about 40, and the first stage by 200 or 300.

As one stands on the ground and looks upon the effects of each of these four stages, spread before the eye in a single view, the comparison of the time relations as represented by the work done almost staggers the imagination, and the impression is very forcibly produced that, if anything, the figures are in error in minimizing the time of the first, second, and third stages and in magnifying the time of the fourth, for the observed activities of the present streams during the present stage are very meager indeed compared to the vast rock erosion of the first stage.

The physical changes in the valleys of the tributary indigenous streams resemble in many ways those of the through-flowing streams. The most marked difference is in the nature of the materials. In most cases these consist of thick beds of fine silts, with occasional deposits of sand, and are of a local origin in all cases except so far as they may be represented by the filling of the mouths of the tributaries along the main rivers with the finer sediments which were being carried through these channels.

During the first stage of erosion of the valleys of the through-flowing streams, all the indigenous streams experienced a rejuvenation, and eroded their valleys to depths varying according to the distance from the through-flowing streams. At the close of this stage nearly all of the drainage had reached a condition in which the floors of the valleys were well graded, and the headwaters had begun active work against the old divides. It was the extensive erosion of this stage, in connection with the dendritic type of the old drainage, which has given a rough and rugged character to so much of the region. An estimate of this erosion, based on some approximate measurements in the field, indicates that between the mean level of the upland plain and the mean level of the present drainage, which is estimated to be about 200 feet, about one-half was removed during this cycle of erosion. This would make a layer of rock about 100 feet in thickness over the entire basin. and gives some idea of the time interval which must be involved. If the rate of 1 foot in 5,000 years would apply to this erosion it would involve a period of 500,000 years.

The change in the conditions of the through-flowing streams which inaugurated the deposition of the gravel trains produced a slack-water stage in all the tributaries. In these slack waters were begun the depositions of the silts, and these continued pari passu with the filling of the valleys of the through-flowing streams, until, at the close of that stage, the tributaries were filled to the level of the highest terraces at their mouths, and aggraded almost, if not quite, to their headwaters.

With the recutting of the channels of the through-flowing streams in their valley deposits began the removal of the silts from all the tributary valleys. During this second cycle of erosion all of the phenomena described in connection with the through-flowing streams, such as deserted channels and minor deflections over rock barriers, occur here also, i. e., the deflection of Mud River over the rock ledge at Barboursville and of the Ohio at St. Marys. It is to be noted, however, that the intermediate terraces along the indigenous streams, though not entirely absent, are far less conspicuous, and it is a comparison of these features which indicates the fact that the erosion of the deposits in the through-flowing stream valleys was accomplished by much larger streams than the present, which probably experienced periodic diminution in volume. The features which characterize the fourth stage in the valleys of the through-flowing streams are also present here, for these indigenous streams are now eroding in places and depositing in others, showing that conditions of equilibrium are not yet fully established, and thus indicating the shortness of the present régime.

THE SLENDER TERRACES.

One of the most marked features of the topography in many sections of this region is the system of horizontal wave terraces which occur very generally distributed throughout the lowland area. They may occur equally as well marked in the plateau region, but they have not been studied there by the writer.

Similar structures have been described by Prof. J. J. Stevenson, Mr. W J McGee, Dr. T. C. Chamberlin, and others, in the basin of the Upper Ohio. Speaking of these as observed in the Upper Ohio region, Dr. Chamberlin says:[a]

"These terraces are far from being constant, and commonly occur only at localities obviously favorable for their formation and preservation. There are usually two, but rarely three or more members, all within a vertical interval seldom exceeding 50 feet. They differ entirely from the preceding classes [the lower river or moraine-headed terraces], being merely slender bands of slope wash lodged along the sides of the hills, reaching a thickness of perhaps 10 feet measured normal to the hill slope, and having sloping surfaces. So far as observed, they maintain a horizontal position. They seem to be just such accumulations as might gather on the margin of a ponded stream in a comparatively limited interval of time. The mean altitude, on both the Ohio and the Monongahela is about 150 feet."

This description will apply almost equally well to the slender horizontal terraces in this middle section of the Ohio Basin. That they were produced by wave action, operating for a short time on a previously graded slope, is clearly shown

---

a Bull. U. S. Geol. Survey No. 58, p. 37.

from a careful study. In general they appear to be similar to a cut made along the hillside for a roadbed. The material removed from the cut is thrown out to make the bank on the lower side. Many measurements show that the volume of the lower embankment is equal to that of the cut, and if replaced would restore the original graded slope of the hill. Above the terrace there is usually a steep face, often of rock, and in many places this rock consists of a soft shale. Only in a few instances were observed some deposits of sand on the surface of the terraces, but in general there are no rounded or sorted materials which would indicate prolonged wave action. The embankment is composed of clay and stone, showing no stratification so far as observed. In several cases there were found well-marked soil lines, extending under the débris of the embankment, which were in accord with the slope of the hillside on which the terraces occurred. Their relation to the stratigraphy, which is often inclined while the terraces are nearly if not quite horizontal, and the character of the cross section, as indicated above, show conclusively that they are not structural or degradational terraces. The lack of a rock platform as a base and the physical character of the material, whose immediate source in the upper cut is evident, shows that they are not remnants of old fluvial gradation plains left in the process of erosion. Pl. XIV, G, shows near the left of the center of the picture a series of three of these terraces in the valley of the Little Muskingum. Pl. VIII, C and D, are nearer views of two of these terraces in the valley of the Raccoon, a few miles west of Gallipolis.

These terraces are rarely found singly, but occur in vertical series of from 3 to 8. They appear to have been produced or preserved in some sections much more than in others, although marks of them are rarely wanting anywhere.

In some basins, like that of Federal Creek, the section of the Hocking between the Athens col and the Federal Creek col, the valley of the Raccoon, and the old Dutch Flats Valley in West Virginia, they are so marked that a map of them would represent a very perfect contour map of the region. They are distributed through a vertical range of about 300 feet, which is considerably more than that reported for the similar terraces in the Upper Ohio Basin. It has been observed in several cases that a line of terraces borders the hills of the valleys on each side of some of the lower cols, which latter show decided marks of having been weir ways at some high water stage, and the terraces occur at the level of the floor of these short channel ways. Again, in a few cases they seem to be associated with a hard resistant bed which has been cut through in the wearing out of some of the old cols by the present streams, and their absence is especially noticeable in the vicinity of these cols. If this observation should be more fully sustained, it would indicate that the terraces were produced by wave action in the impounded waters during a short period of retardation in the cutting out of the old cols.

They are found on the walls of the old valley and on the valley walls of the present streams, so that their date of origin, at least of the lower ones, must be subsequent to the erosion of the present deep valleys. It seems probable that they all belong to the same period, and that that period was very recent.

THE GLACIAL DEPOSITS.

As the basin lies entirely beyond the glaciated area, there are within it no deposits directly attributable to the action of land ice. The glacial deposits consist of two classes; first, the gravel trains, which extend throughout the valleys of the through-flowing streams. These are discussed under the head of the deposits on the floors of the present valleys. The second class includes the scattered stones and bowlders of foreign rocks, which have often been reported in the past from various parts of the basin. They are of scientific interest principally from their horizontal and vertical distribution. There is scarcely any large area within the lowland section from which they have not been reported, and at elevations varying from the present drainage to 300 feet above the Ohio River near Syracuse, where the author has collected many scattered stones at that elevation. They range in size from small pebbles to bowlders of from 1 to 2 feet diameter. It seems evident from the data in hand that the suggestion of many of the earlier writers that these must be attributed to floating ice during a stage of submergence, when the waters covered almost the entire region to an elevation of about 700 to 900 feet above tide, must be correct As far as the observations of the writer extend, these scattered erratics do not occur above the level of the highest slender terraces.

HISTORICAL RELATIONS OF THE OLD HIGH-LEVEL DRAINAGE SYSTEM TO THE PRESENT.

In tracing the sequence of events, as indicated by the foregoing observations and deductions, from the time when the drainage of this region occupied the horizontal and vertical positions indicated by the restoration to the present drainage conditions, it is very evident that it is necessary to consider the operation of all the forces involved and the relations to the sequence of events in the circumjacent regions. The phenomena of changes in drainage have been very thoroughly studied within the past few years, and most of the processes by which changes are brought about are well understood. For convenience of reference these processes may be divided, according to the principal factor operative in producing the change, into two classes. The first may be called the normal and the second catastrophic, and the latter may easily be divided into the internal and superficial. Under the first class belong those processes which are normally present in the action of the streams, such as the erosion of channels, aggradation, migration of divides, piracy, etc. Under the second class are included such processes as are produced by volcanic and earthquake phenomena and by the formation of ice masses, avalanches, landslides, etc. The first minor division of this class would include the volcanic activities, lava flows, and earthquake phenomena, with the associated change of level, fissuring, and faulting, and the second division would include the damming action produced by the avalanche and the landslide, and by ice gorging and all glacial phenomena.

In almost every modification in drainage it is probable that many of these processes work in very intimate combination. It is certain that the normal activities of the running water play an important part in every readjustment, although the major factor may belong to the catastrophic group.

A complete discussion of this subject will not here be undertaken, but, by way of introduction to an interpretation of the history of the region under investigation, attention will be called to a few of the principles which will be considered in this connection. Among the first class, besides the action of the stream under the influence of deformations of its bed and the discovery of rock structures producing subsequent modification, there are a group of activities associated with aggradation which have not been very fully discussed, although the principles involved are well known.

When the grade of a stream is reduced, either by the placing of an obstruction across its course or by a deformation which may affect a part or all of its length, the stream at once loses in its carrying power. If the stream is well graded and a deformation reduces the grade throughout its entire length, it will at once become a depositing stream from its mouth to its headwaters. If a barrier is raised across its lower course the slack-water condition will extend itself headward at a rate dependent upon the supply of the material. It will begin its deposits in the lower slack-water section, and the aggradation will from this point be extended toward the headwaters. In this extension the field of maximum deposition will be at the point where the change in the rate of flow is first met, as here the velocity of the current will be checked and its load will be dropped. This process will continue until a new grade is established throughout the course of the stream. If it should happen that the obstruction were placed suddenly across the stream's course, and were of sufficient size, the slack-water conditions might extend for a considerable distance up the valley, in which case the sediments would accumulate at some point very remote from the obstruction, while between this point and the newly established point of overflow there would be but a small amount of sedimentation. Should it happen that the height of the barrier were increased periodically, the slack-water conditions would be carried at intervals farther up the valley, and thus thicker deposits would accumulate at several points throughout the valley than in the intermediate spaces. It is evident that during such conditions of deposition and aggradation the main line of the current action would become clogged, and that the stream would raise its floor above that of its surrounding flood plain; its channel would thus be deflected over other parts of the plain. These processes are abundantly illustrated in every aggrading stream. If the streams are in somewhat narrow valleys and the topographic features of the region through which they pass are of a mature form, and the cols at the headwaters of the minor tributaries of contiguous basins are very low, the accumulation of the silt deposits along the main valley may be of sufficient magnitude to deflect the streams over the low cols into adjacent basins, or perhaps over low cols separating the tributary waters of the same basin. In either case it is evident that as soon as the waters rise high enough to overflow the col they will find on the opposite side a sharp grade, which represents the difference in level between the waters above the obstruction and the waters below the obstruction which would be backed up in the valley on the opposite side of the col. As the superficial portions of this part of the divide are, from the nature of the case, usually very incoherent, and disintegration is usually quite extensive, it is evident that the stream will at once rapidly cut out the upper soil portion of the col, and in so doing will desert the portion of the valley which is occupied by the obstruction which it has itself built up. If it finds in its new course a free circulation and an increased gradient, it will never return to its old channel, but if the new course is only a temporary relief it may also become clogged with deposits and the stream may rise to the level of its former obstructions and return to its former course, or it may discover still another new route.

From the fact that the line of maximum deposition in an aggrading stream will be in the direct line of current action, the stream ever tending to build up an obstruction in its own path, it follows that the new course which it must discover will be more circuitous than the one which it will abandon.

Again, in the case of eroding streams, if two courses are open to the stream, the tendency will be for the stream to eventually gain the shorter course if the resistance to its cutting is about the same along each line, as the shorter course will have the higher gradient and hence the greater velocity. This difference in velocity may be great enough in some cases to determine the location of the stream along the short line, even when the resistance to be overcome is greater. The extreme is found when the stream may be so balanced that it will deposit along the longer route while it is removing a barrier in its shorter course, and eventually the longer route will be deserted.

Under the second class of processes—the catastrophic—the modifications are usually produced suddenly and are largely independent of the normal activities of the streams. The internal and superficial agents operate in very similar ways; by the actions of lava flows, earthquakes, faultings, landslides, and glacial ice the streams are suddenly dammed up and caused to seek new lines, or the opening of fissures determine the location of the new stream courses. The competency of land ice to permanently dam a river course and cause the deflection of the stream has been questioned by some writers, but there are abundant observations which show that such conditions actually occur at the present time in connection with existing ice masses. The number of cases which have been described, in which the data show almost to a demonstration that the great continental glacier produced similar results, places this question beyond a doubt. The glacier has been shown to have operated in producing modifications by the obstruction of the valleys with its deposits, and also directly by its own mass. Where the mass of the ice forms the dam it is evident that the ice must remain during the entire time of the cutting out of the restraining col to a point below the base of the floor upon which the ice itself rests. A subsequent removal of the ice by melting will then leave the stream in the new course, as it has cut so deep that it can not regain its old channel after the removal of the obstruction. Where morainic material forms the obstruction it is evident that the deposits must exceed the elevation of the lowest col in the rim of the basin above the dam. Examples of both of these types are very numerous and are fully described by many writers.

It is the application of these principles, with others, to the sequence of events, which is indicated by the field studies, which must furnish the interpretation of the drainage modifications in this region. This sequence of events may be briefly stated as follows:

1. The reduction of the region to the base-level plain of the present upper Tertiary peneplain by a very remote and extensive cycle of erosion.

2. The elevation of this peneplain, with a rejuvenation of its matured drainage and the entrenchment of all its streams into deep valleys which retain the extensive meanderings of the ancestral streams.

3. The complete gradation of the drainage with the development of mature topographic features and the reduction of the old peneplain to such an extent that it is only represented by the even line of numerous peaks and long continuous ridges, and the cutting of these ridges at many points by the headwater streams, with the formation of very numerous low cols, and the development of a very perfect type of dendritic drainage.

4. The production of slack-water conditions throughout the lowland area and extending well up into the plateau region, with a widespread deposition of silts, to depths of as much as 50 to 60 feet at certain points, and numerous deflections of drainage from the old lines.

5. A rejuvenation, accompanied by a very deep erosion of the new lines of drainage extending below the level of the old system, with the introduction of tor-

rential waters at four points in the rim of the basin (New Martinsville on the Ohio, Bluerock on the Muskingum, Lick Run on the Hocking, and Chillicothe on the Scioto) with egress at one point (Manchester on the Ohio), and the complete accommodation of the minor drainage lines to this newly established low level.

6. The filling of the valleys of the through-flowing streams with glacial gravel trains and the silting of all of the tributaries to corresponding levels.

7. The erosion of the valley fillings and the cutting of the intermediate terraces to the depths of the present streams.

8. A high-water flood stage in which the whole region is submerged to great depths with the production of the slender terraces and the scattering of the glacial pebbles and bowlders by floating ice, the periodic falling off of these waters producing the series of slender terraces.

9. The reduction of the volume of the streams and the continued stream action to the present time under present conditions.

It is evident that the first step in the order of things which led to the change in the drainage appears in the fourth stage in the production of slack-water conditions which resulted in the widespread deposition of the silts on the old valley floors, and this in greatest abundance along the main lines. The fluviatile rather than lacustrine origin of these silts is shown both by their structure and their general distribution. They are spread over the floors of the old valleys as a rather even sheet, occupying the full widths of the valleys, except where evidently removed by subsequent erosion, but they are not observed to lie upon the side slopes of the valleys above the valley filling, showing that the water in which they were laid down was not more extensive than the present limits of the deposits, for had the waters been deep and had the deposits been left at various elevations on the slopes of the valleys it is scarcely possible, with the very abundant opportunities for preservation on protected platforms, that all of the upper deposits should have been completely removed to the level of the old floors. As the upper surface of these deposits now descend from the rim of the basin toward its axial line, it must be assumed that if they were laid down in a lake the lake must have been of considerable depth in the center of the basin, or else that the basin had suffered a very marked and peculiar deformation, of which there is no evidence. Therefore, as the deposits seem to have been made in rather shallow water of no greater extent than the upper surface of the deposits, it follows that the waters must have been slow-moving currents rather than lacustrine.

Again, it is common to find scattered through these silts coarser bands of sandy material, such as would not be carried in suspension, but such as would indicate a certain amount of current action.

Perhaps the most conclusive evidence of the fluviatile origin of these silts is found in their distribution. They occur along the courses of the old valleys and appear in great thickness along the larger streams. This would be the normal distribution if they were laid down by aggrading streams, but, on the other hand, if they were deposited by streams discharging into a lake they would be distributed in delta-like forms in the headwater sections of the submerged valleys. The main lines would be the last to become obstructed. The two forms of deposition would produce very different effects upon the distribution of the subsequent drainage. The predominating tendency in the case of an aggrading stream is toward a shifting of the position of the stream line, while in the case of the filling and draining of a lake basin the predominating tendency of the resulting drainage would be to follow the deeper line of the basin, and in such a lake as might be assumed to have occupied this region— which must have been scarcely more than a drowned-river system very much like that of the Chesapeake Bay—the new drainage would conform almost exactly to the old lines

Apparently closely associated with this silting are the many and irregular drainage modifications, suggesting at once their relations to the conditions of aggradation in the old systems. The fluviatile origin of these deposits and the fact that they indicate a slack-water condition of the drainage at this time seems fairly well established.

This leads to the question as to the cause of the slack-water conditions. One of the most potent factors in producing slack-water conditions is a deformation of the surface. If this factor is applied to the case in hand it finds no direct support. The modifications are so numerous and their distribution is so extremely irregular as to make this inherently improbable. Besides, the old floor constitutes now a perfectly graded system (when the old drainage is restored), so that, to produce these drainage modifications by differential warpings of the floor of the basin and then restore the present condition, it would be necessary to assume that after the modifications were well established an exactly opposite warping took place. Such an exact duplication of a complicated movement in exactly the reverse order seems extremely improbable and finds no support in the observed facts. On account of the great diversity in the direction of modifications it is impossible to assume that they were produced by a uniform tilting of the basin in any particular direction. The only possible condition which might be attributed to this cause is that in which there was a general differential subsidence of the highest parts of the several basins with such a motion that all plains remained nearly parallel to their former positions. While this explanation might account for the observed silting by producing the slack-water conditions within the basin, there are no observed data in support of this hypothesis except the presence of the silts themselves.

As the modifications within this basin are closely associated with the introduction of the torrential waters from the adjacent basins to the northeast, it is evident that the causal relations will also be intimately connected.   It seems necessary, therefore, to consider the history of events in these regions in this connection, as there appear here only the facts of aggradation, without a sufficient suggestion of the cause.

It has been clearly shown by the authors already referred to that in the Upper ●hio Basin the modifications of drainage were intimately associated with the glacial phenomena of the Pleistocene, and that the introduction of the drainage of that basin into the one under consideration, across the New Martinsville col, is to be directly attributed to the accumulation of glacial ice and drift in the mouths of the old valleys, which originally discharged northward, against the direction of the ice movement.

The writer has shown that the deflection of the waters of the upper Muskingum and Tuscarawas basins from the Newark Valley over the Bluerock col was caused by the occupation of the old Newark Valley by the ice and by the great deposit of glacial material left in the old valley at Hanover, while the deflection of the head-waters of the Hocking over the Lick Run col was undoubtedly produced by the direct action of the ice mass.   A similar deflection on Paint Creek, produced by the ice itself, is described by Mr. Fowke, and one at the Beech Flats is described by the writer.[a]   The action of the ice and of its morainic deposits in reversing the headwaters of the North Fork of Salt Creek at Adelphi has been referred to in detail in this paper.

From these considerations the question naturally arises, Did the ice have anything to do with the deflections in this basin; and if so, to what extent?

As the ice front did not invade the basin, and as there are no morainic deposits present, it is evident that if the ice were a factor in producing the modifications it must have operated indirectly.

It is to be observed that the main axis of drainage of this basin is almost normal to the line marking the limits of the ice at its maximum stage of advance, and that the direction of discharge is directly opposed to the direction of flow of the ice (see map, Pl. I); it is evident, then, that the ice stood as a most effective barrier to the northwestward discharge of the waters of this basin.   The encroachment of the advancing ice front on the old outlet of the basin furnishes the needed factor to account for the inauguration of the slack-water conditions.

It has been shown by Dr. Chamberlin and others that the mass of the ice itself, by its own attraction, would have a marked effect upon the water levels for many miles.   A complete discussion of these attractive influences is given by Dr. Chamberlin, in collaboration with Prof. R. D. Salisbury, in a description of the loess in the Mississippi Valley.[b]   From this discussion it appears that this attraction would amount to about 4 inches to the mile along a line normal to the ice front within a

---

[a] Bull. Sci. Lab. Denison University, Vol. IX, Part I.
[b] Preliminary paper on the driftless area of the Upper Mississippi Valley, Sixth Ann. Rept. U. S. Geol. Survey, p. 291.

distance of one degree. If the old river were perfectly graded, it is evident that this effect alone would extend far into the region of the silt deposits. On the principles previously stated, any effect on the lower waters of the graded streams would be extended headward until a new grade was established.

As the advance of the ice was very slow, it is evident that conditions favorable for the aggradation of the opposed stream were presented. This is equivalent to saying that the barrier increased in altitude slowly, although not necessarily at a regular rate. From our knowledge of the action of water currents along the margins of ice masses, it is evident that the stream did not yield its ground without a struggle. As the waters became imponded they sought out new lines of discharge, circumventing the advancing ice front or cutting tunnel ways under its border. With the closing of each new outlet the waters were more impounded and rose to higher levels, and the slack-water effects were extended farther toward the headwaters of the basin. The last step in this process seems to have been taken when the ice, after many struggles with the contending waters, reached a position in its advance which carried the waters of this basin high enough to flow over the low col in the divide near Manchester.

We find, then, in the relations of the drainage to the known position of the ice front an all-sufficient cause to account for the production of the slack-water conditions, and their prolonged duration and gradual extension throughout the basin.

With the data already in hand it seems possible to trace the larger features of the sequence of events throughout this fourth stage with considerable certainty. It is this sequence which shows the relations between the old and new systems, and how the latter has been derived from the former. If the suggestions offered proved later not to be the exact order, it is believed that they will at least indicate the broad line along which the future interpretation must be worked out by careful detailed examinations of the whole area involved.

With the first obstruction of the drainage began the deposition of the silts in the lower section of the old valley. These deposits were extended up the axial stream along the line of the old Teays River and increased in thickness until an accumulation of from 20 to 30 feet was laid down on the floor of the old Flatwoods Valley, and of over 50 feet in the upper portion of Teays Valley. From the excessive accumulation at these points it may be assumed that they represented points where the slack-water conditions were first encountered by the stream for a considerable interval, which probably represented rather uniform conditions at the barrier below. By the accumulations on the floor of the Flatwoods Valley (Pl. IX) the old stream was deflected to the north side of the deposits and meandered around the hills now found between the old valley and the present Ohio. Accompanying these excessive accumulations in the axial valley there was extensive aggradation also along the northern branch, and the accumulations at Camba amounted to over 90 feet (Pl. XIII).   This also was probably a nodal point on the stream from which the aggradation was more perfect eastward, extending up the valley of the old Marietta River.   This accumulation deflected the waters across the divide between the old Point Pleasant River and the old northward-flowing Little Guyandot Creek.   In circulating among the hills in this divide they crossed several low cols, giving the grouped distribution of the cols indicated in the vicinity of Gallipolis.   Being unable to follow down the old Little Guyandot Creek on account of the same obstructions in the lower course at Camba, the waters found an easier route by following the course of the old Little Guyandot to the mouth of the present stream and turning westward up a small tributary across the old Crown City col and running down the tributary on the opposite side to the main axial stream at Huntington.   This route is about as long from the point of divergence at Point Pleasant to the axial stream at Huntington as its older one from Point Pleasant to Beaver, but it met the axial stream at a somewhat higher level at its new point of confluence.

Probably almost simultaneously with the changes just described the deposits in Teays Valley had accumulated to such a depth as to deflect the upper waters of the axial stream over the low cols in the divide at Winfield and on the Hurricane, the included area probably existing as an island (Pl. XIII).   These waters followed down the old Point Pleasant River, but near the mouth of this stream they began to feel the effects of the aggradation produced by the obstruction at Camba, which resulted in the depositions in the old valley at the mouth of the stream, and hence the waters cut across the col at Point Pleasant and joined those of the old Marietta River at that point, thus returning again to the axial valley at Huntington, having found this route around the obstruction built up in the upper section of Teays Valley.   It is possible that at this same time a minor current may have set along the Raccoon and Symmes Creek valleys over the Marion col.

Another possible alternative interpretation for these larger deflections might indicate that the old drainage was first obstructed by the accumulation in Teays Valley and that the axial stream was thus deflected over the Winfield col to the waters of the old Marietta River, that for a time the upper waters followed that stream to the outlet, and that the deposition of the accumulation at Camba may have occurred a little later, so that the deflection of the waters across the Crown City col came subsequently. Between these two suggestions there seems to be but little choice as far as the data in hand indicate.   The facility with which these streams crossed the old Marion-Crown City-Hurricane-Winfield divide is manifest from a study of the restored drainage as represented on the map (Pl. I), as it is evident that this divide must have migrated very extensively during the development of the old drainage from the old

Marietta River toward the old Teays River, and the tributary streams of the former had extended their headwaters very close to the old axial valley in the region along the present Teays Valley, so that at numerous points this divide presented very low cols; and the fact that the Kanawha has crossed the divide in one direction and the Ohio in the opposite direction suggests the close balance which existed between the two major lines of drainage in the old basin.

The deposits on the floor of the lower portion of the old Albany River (Pl. XVI) resulted in the deflection of the drainage of this part of its basin over the Minerton col along the line of the present Raccoon, while the deposits in the old valley south of Zalaski caused the waters of the valley above to be deflected over the old Moonville col and into another tributary of Albany River. The excessive accumulations on the valley floor at Albany deflected the upper waters of that old river eastward across the Athens col and thence along the line of that section of the present Hocking. The accumulations on the old Dutch Flats Valley in West Virginia (see Pl. XIII) caused the old Marietta River to wander around among the low hills in the vicinity of Pomeroy, where it passed by another group of cols very similar to those in the vicinity of Gallipolis. The accumulations on the valley floors at Wilkesville and Salem Center deflected the waters of Leading Creek eastward across the Langsville col to the newly forming main line. The accumulations along the floor of the old valley between the Ohio at Hartford and Hockingport caused the Marietta River in its upper stretches to meander around among the hills lying east of the old channel, but it found this channel again at Hartford, this deflection being similar to that at Ashland and Ironton from the Flatwoods Valley, but on a larger scale, while the accumulations in the valley south of Guysville turned the waters of this section of the Hocking across the Federal Creek col and along the small streams occupying the present lower course of the valley. At Layman and Barlow (Pl. XI) the silting was sufficient to deflect the waters of Wolf Creek and of the stream occupying the present portion of the Muskingum Valley below Bluerock across the Meigs Creek and Lowell cols to the old river line at Marietta.

Further details connected with many of the minor modifications described might be presented, but it is thought that sufficient has been said to indicate the action of the aggrading streams in deflecting their courses around the deposits which were laid down in the more direct line. It will be observed that, in all cases referred to, the new line is much more circuitous than the old line, and, as was suggested in considering the action of aggrading streams, this would be the normal type of modification. That many of the changes here indicated as occurring in succession were in reality almost simultaneous seems quite probable from the fact that the accumulation of the old silts in the valleys would take place very generally over the basin at the same time by the aggrading action of all of the streams. With such close balances between the various elements it would be extremely difficult to determine absolutely the succession of changes.

The next step in the history of the region appears to be the fixing, as it were, of all of these modifications and the development of the present system. The manner in which this was accomplished is the next question to be solved. We may readily postulate such an elevation of the basin as would produce a rejuvenation of the streams and increased erosion, but it seems that there is a more natural explanation—one which is supported by the facts of observation and the relations to the existing conditions. It is evident that as long as the waters of the old system escaped along the edge of the ice the torrents supplied from its melting would join the bordering stream, but as soon as this outlet was stopped and the impounded waters crossed the Manchester col all the waters from the melting ice above the point where the barrier was produced would be added to the current flowing over this outlet. This current extended down the present Scioto Valley and along the line of the present Ohio to Manchester. The altitude of the Manchester col, 'or possibly of some other col still farther down the Ohio, would represent the maximum elevation to which the waters would rise within the basin. This would determine the upper limit of the silting. With the beginning of the erosion of this col the fourth stage was terminated and the fifth inaugurated; for, as the elevation of the drainage levels was produced by the ice, it seems reasonable to suppose that when the waters succeeded in permanently escaping from the action of the ice barrier they would begin at once to erode their valleys toward the former level, and the permanent location of the streams would depend on their particular location at the time of the change in the conditions from aggradation to erosion. As the Manchester col was probably worn away with considerable rapidity, on account of the great volume of water which crossed it and its silty character, the deepening of the channels in the old silt deposits would progress rapidly toward the headwaters of the basin. With the reversal of the drainage along the line of the Scioto it appears that the main current from the drainage of the basin took the shorter route across the Portsmouth col rather than the longer route by the way of Waverly and down the Scioto.

During the early stages of this erosion the minor drainage channels which were associated as tributaries to the old abandoned valleys wandered in various directions through and across the old valleys in their endeavors to find the easiest route to the rapidly developing new drainage. The minor irregularities in the depths of the silt fillings on the old valleys were important factors in determining these lesser modifications and deflections. The complex arrangement of the smaller streams which now cross the old valleys in apparent obedience to no law or order, such as those in the vicinity of Wellston, Hamden Junction, the Dutch Flats Valley, and the Hurricane in Teays Valley, finds a very rational explanation well borne out by observed phenomena. The amount of erosion which would be accomplished during this stage would depend on several factors. If the newly found outlet had a longer route to the sea than the old drainage, it is evident that the erosion would reach a base-level condition at an altitude somewhat above its old level, but if the new route over the Manchester col was a shorter course to the sea—and many facts in hand point strongly in this direction—the new base-level of erosion would lie below the old one. This factor would also be influenced by any movement which the basin may have suffered during the preceding or present stages. If the movement were in the form of a depression, the effect would be the same as in the case in which the new outlet was the longer route to the sea, while if the movement were that of an elevation the effect would be the same as in the case of the shorter route to the sea. Reasons have already been presented for considering the hypothesis of an extended movement very doubtful.

Another factor which would influence the extent of this erosion is the volume of the water. This must have been greatly increased by the addition of the water resulting from the melting of the superficial surface of the front of the ice. The result of the combination of these forces is evident in the deep erosion of the newly formed drainage. From the great extent of this erosion, amounting to about 150 feet below the level of the old drainage, it seems evident that either the new route was very much shorter to the sea or that the basin experienced during this stage some elevation. To the extent that one of these factors is increased, the other is correspondingly decreased, and at present there seems no means of determining the quantitative value of each. With the rejuvenation of the drainage the streams which followed the old lines had to work entirely in the silt deposits of the preceding stage, but where in their meanderings they had crossed the old watersheds or had been deflected toward the sides of the old valleys against the bordering hills they soon discovered the underlying rock and began its excavation. The rate of cutting of the new system would therefore be determined by the rate of erosion at the points where the streams discovered these rocks barriers.

As has been already suggested, the topographic features at these points indicate that this rate of erosion was extremely rapid. To what extent this had progressed before the inauguration of the sixth stage, which was marked by the introduction of the torrential waters at the Lick Run, Bluerock, and New Martinsville cols, seems very difficult to determine. This much, however, seems certain: That the erosion had deepened the new channels so far below the level of the old valley floors that the increased volume of water produced by the torrential floods which came down the valleys of these through-flowing rivers did not rise high enough to overflow the floors of the old valleys adjacent to their courses; and it is also a fact that they followed most closely the lines established by the meanderings of the streams during the fourth stage; for instance, when the flood waters were poured over the Bluerock col (see Pl. XI), had they been of sufficient volume to rise over the valley floor at Layman it would seem that they would have more easily eroded a channel through the soft silt deposits on that floor than to have taken the longer route and eroded the rock barrier at the Meigs Creek col. The same is true with reference to the flood waters poured down the Hocking Valley into the old Albany River, the line of that river past Albany to the valley of the Raccoon being much more direct than the line across the Athens col and so down to the Ohio. It is hardly possible that the flood waters should ever have occupied these old valley floors without having left some evidence of that occupancy. The evidence in support of this conclusion is even more convincing when the course of the Ohio in relation to the remnants of the old valley floors is considered, especially in the region of the old Dutch Flats Valley and the Flatwoods Valley, neither of which shows the slightest evidence of ever having been occupied by the flood waters of the Ohio. It seems possible to suppose, however, that the waters of these through-flowing streams were introduced into the basin at some time before the final close of this period of erosion, for their channels are deepened, at the points where they cross the old cols, to an extent corresponding to the gradation plains represented by the rock floors of the present drainage. These waters, therefore, may possibly have assisted in the deep erosion of the channels of the through-flowing streams below the level of the old valley floors. It is certain that they occupied these valleys at the grades represented by the present rock floors before the introduction of the glacial materials which subsequently filled these valleys with the gravel trains.

The exact order in which these streams were introduced into the basin seems difficult to determine, but this question may be decided by more detailed field work. It is quite evident, however, that the reversal of the Scioto preceded the introduction of the waters of the Hocking, the Muskingum, and the Ohio, by the time interval represented by the amount of erosion of the present streams to a little below the level of the old valley floors. It thus appears that the reversal of the Scioto took place at the very beginning of the period of erosion, and the introduction of other through-flowing streams somewhere near the middle or a little past the middle of that period. Some support for this view is found also in the somewhat older characters of the Scioto Valley.

This cycle of erosion was terminated by the beginning of the deposition of the gravels, sands, and silts in the valleys of the through-flowing streams, which marks the inauguration of the sixth stage. Writers on this subject have usually attributed this change of conditions to a general depression which produced a slack-water drainage and converted the streams into aggrading rather than eroding streams. This assumes that the eroding streams were carrying the load of débris, and that the reduction in the grade caused them to drop this load along the valleys. The evidence in the deposits themselves indicates that the drainage was very vigorous, and would seem to oppose any hypothesis which would indicate slack-water conditions. The vigorous drainage thus indicated would seem to point to a high rather than low altitude of the land; at the same time it would seem more reasonable to suppose that the beginning of the deposition of the gravels was due rather to the overloading of the streams which were previously doing the work of erosion, and that this stage of aggradation was the result of this overloading rather than of a general depression. This conforms more closely to the conditions indicated by the deposits. The suggestion is therefore offered that the deposition of the gravel trains was the result of the overloading of the streams at the time of the maximum advance of the ice when its border had so encroached upon the region that its waters were confined to these few lines of discharge, and the material which otherwise would be distributed in a great morainic fan was concentrated along these lines. The termination of this stage was therefore brought about by the beginning of the recession of the ice which supplied the gravels, for with the beginning of the recession of the ice the materials supplied by it would be distributed over the deserted floor of the glacier in its recession and less completely supplied to the streams which discharged from its receding front. It is evident that under such conditions the volume of the water would not decrease at the same rate as the decrease in the supply of material, and it would follow in consequence that the streams which were aggrading under the heavy load of material would, when relieved of that load, begin at once to degrade and remove the obstructions produced by previous depositions.

With this change in the relations of load to current action was inaugurated the seventh stage in the history of events. The apparent periodic falling off in the volume of the water during the seventh stage, which is represented by the production of the intermediate terraces, finds a sufficient explanation in the conditions accompanying the recession of the ice. With that recession and the opening of new lines of discharge for the glacial waters the volumes of the through-flowing streams would be periodically reduced, until at last the final withdrawing of the ice waters beyond the last weirway which opened into this basin would reduce the volume of all of the streams to the conditions expressed by the meteoric waters alone, and with this last step in the reduction of the waters—the final withdrawal of the glacial current—the seventh stage was terminated.[a]

The ninth stage in the cycle extends from the close of the seventh stage to the present, so far as fluvial work is concerned, and is represented by the observed activities of the present streams.

---

[a] See also the causes of transformation set forth by Dr. Chamberlin in Mon. U. S. Geol. Survey Vol. XXV, p. 251.

The exact relation of the eighth stage, which is lacustrine, as indicated in the outline, is not so evident. Its place may possibly be at the very close of the sixth instead of the seventh stage, but certainly it must be subsequent to the deep cutting of the present valleys. The lake conditions, which are represented by the slender terraces or wave beaches formed during the stage, seem to be associated with an obstruction to the drainage beyond the limits of this basin. This subject needs much further investigation. From the work of Prof. Joseph F. James and Mr. Gerard Fowke it seems evident that the sequence of events associated with the drainage modifications in the Cincinnati region must have had an influence upon the history of the regions farther up the river. Mr. Fowke has shown that the old drainage system in this region, which is undoubtedly to be correlated with the old drainage of the basin under discussion, originally discharged northward, while Professor James has shown that the course of the Ohio, corresponding to the deepest cycle of erosion represented by the valley floors at the close of the fifth stage as above outlined, was up the present Mill Creek Valley to Hamilton and thence down the Great Miami to Lawrenceport, Ind., and that the channel of the Ohio immediately below the city is of more recent date. This relation seems to suggest that the blocking of this channel-way by the extensive glacial deposits around Hamilton, subsequent to the cutting of the deep gorge, raised the waters of the upper basin to the elevation of the col immediately below the city.

The structure of the bounding hills—alternate beds of soft and hard limestone—seems to suggest that the cutting of this col would be marked by intervals of rapid erosion alternating with intervals of rather slow cutting, while the time involved in the entire removal of the obstruction would be comparatively brief. The stages of slow cutting may be found to correspond with the brief stationary conditions of the water level in the upper lake portion permitting the formation of the wave-marked slender terraces.[a] It must be distinctly borne in mind that this explanation has no relation to the Cincinnati ice-dam theory as proposed by Prof. G. F. Wright, for these terraces are not discussed in connection with that theory, except so far as suggested by Dr. Chamberlin in his reference to similar structures in the Upper Ohio Basin.[b]

While it may be shown that the ice at some stage extended beyond the present Ohio near Cincinnati, it does not follow that the present valley was previously developed; in fact, it seems very improbable from the data in hand. The ice-dam theory presupposes the existence of the present valley, and it is largely based on the phenomena associated with the old high-level valley floors and gradation plains, which are in this article, it is thought, rather conclusively shown to have no relation what-

[a] See also the "stoping" process, by Chamberlin, Mon. U. S. Geol. Survey Vol. XXV, p. 351.
[b] Introduction to Bull. U. S. Geol. Survey No. 58, p. 38, sec. 4 of his summary.

ever to the supposed obstruction at Cincinnati, but to form an integral part of a normal sequence of events, and to be in no sense catastrophic in character.

It is also to be noted that while the writer does not deny, but rather admits, the competency of ice to obstruct drainage lines under certain conditions, the retention of the waters during this stage of submergence, in which the slender terraces were produced and glacial stones were scattered over the highlands of the region, is not attributed to an ice barrier, but to the resistance offered to the cutting of an old divide. The ice and its deposits had completely obstructed the former line of discharge at a more remote point. It seems highly probable that in this case the beginning of this rock cutting may have been by the action of a stream working under the attenuated edge of the ice when it was at its stage of maximum advance on the Kentucky hills, and hence the gorge produced by the cutting lies within the area marked by the presence of the ice. Even such a rock barrier as is here proposed could not, in the mind of the writer, have furnished the necessary conditions to produce the extensive phenomena attributed to the much less effective ice barrier proposed by the ice-dam theory.

The final question to be considered is the time relations of the various stages in this cycle and their correlation to the recognized phenomena in the circumjacent regions. The time involved in the erosion represented by the first stage in the production of the old peneplain is very great, and probably expresses the sum of all of the erosion cycles from the beginning of the Permian to some point in the Tertiary.[a]

The inauguration of the second stage by the elevation of the old peneplain may be considered as occurring somewhere in the Eocene, and the continuation of that cycle through the third stage, resulting in the production of the mature topography and the almost perfect gradation of all of the drainage, is represented by the time interval from the elevation of the peneplain to the beginning of the fourth stage; that is, to the production of the slack-water conditions. If our interpretation is correct, that these slack-water conditions were produced by the invasion of the ice into the lower section of the drainage lines coming from this basin, it fixes the date of the beginning of this cycle at some point in the Pleistocene. If the gravel trains which fill the deeply cut valleys of the through-flowing streams are correctly correlated with the morainic terraces of the most recent glacial epoch, it at once becomes evident that the inauguration of the slack-water conditions at the beginning of the fourth stage is separated from the valley filling of the sixth stage by the time represented by the amount of silting in the fourth stage and the amount of erosion represented by the entire interval of the fifth stage. The question involves, then, the consideration of these two intervals; that is, the interval of silt

---

[a] The time correlation is open to much question and is not improbably later than here assigned.—T. C. C.

deposition and the interval of subsequent valley erosion. There seems to be at hand no convenient means of measuring the rate of accumulation of the old silts in the old valleys during the fourth stage, for the conditions under which these silts were deposited seem to have been somewhat exceptional. The character of the silts themselves, however, indicates a rather slow rate of deposition, and when it is considered that along the principal lines this accumulation amounts to from 60 to possibly 100 feet in thickness, it suggests that the interval was one of great duration. It may be assumed that the time involved in this interval is also represented by the time at which the ice first encroached upon the basin and the time of its maximum advance and the deflection of the drainage over the Manchester col. While it has not been clearly shown as yet where the ice first encroached upon this old drainage line, because the latter has not as yet been located with certainty, the evidence in hand seems to indicate that the obstruction may have first been placed across the drainage many miles from the point of maximum advance of the ice. In view of this it would appear that the time interval represented by the deposition of the silts might also be expressed in terms of the rate of advance of the ice, but as this is largely a matter of speculation it does not furnish any satisfactory data for an exact determination of the length of this interval.

The interval represented by the erosion of the fifth stage offers somewhat more favorable conditions for a basis of calculations, although even here only general and approximate results can be obtained at the present time. The variability of the factors involved in this erosion causes very great uncertainty in any calculation which attempts to approach accuracy, but the general facts may be indicative of certain comparative relations of the time intervals.

The total amount of erosion accomplished extends through a vertical range of about 350 feet, of which the upper 150 feet is not represented by the erosion of the valleys through a continuous rock channel, but by the cutting of the old cols crossed in the development of the present drainage and the removal of the silts deposited along the lines of the old valleys which the new drainage occupies. This cutting was done by streams having volumes comparable to the indigenous streams of the region. The cutting of the lower 200 feet of this section was almost entirely through continuous rock channels, but in the case of the through-flowing streams the torrential waters may have been an important factor in determining the rate of this cutting. While it is true that all of the indigenous drainage accomplished sufficient erosion to form well-graded slopes throughout most of the courses of the streams, it is to be noted that if the through-flowing streams cut the deep gorges of the Ohio, the Muskingum, the Hocking, and the Scioto, all of the minor tributary drainage would thereby discover the most favorable conditions for extremely rapid erosion, and this is manifest in the sharp slopes which are everywhere present, representing the grade slopes of this cycle.   But, with the most conservative estimates, the immense amount of work accomplished during this interval points with no uncertain finger to a very great lapse of time.

An attempt has been made to estimate this time also from the amount of leaching and weathering of the silt deposits on the floors of the old deserted valleys, and the conclusions arrived at are very greatly different from those deduced by some other observers.   The occurrence of buried wood and organic matter in these silts and on the floors of the old valleys beneath the silts has been affirmed to indicate the very recent origin of these deposits, but the author can see no reason why organic matter under such favorable conditions for preservation as are presented by burial under these silts and in the presence of constant moisture conditions might not remain for untold ages with very little, if any, change.   At the same time the proof of the very great age of these deposits, which seems to have been entirely overlooked by the earlier observers, appears in the amount of leaching and weathering which these deposits have suffered.   This has already been indicated as reaching to depths of from 30 to 40 feet below the surface, where the old buried gravels are found to be completely decayed, and within extreme depths from the surface every particle of soluble matter has been leached out, and the upper superficial layers so completely acted upon by the percolating water and the disintegrating effects of the atmosphere that the lands are everywhere recognized by their white and light-yellow color and lack of fertility.

In comparison with the amount of weathering represented by the clays and silts of the last ice invasion to the north of this district, the time interval represented by the leaching of these old valley silts would seem to be many fold that of the time since the deposition of the most recent glacial deposits.

These facts seem to indicate that the beginning of the slack-water conditions represented by the fourth stage must be located at the time of the earliest advent of the ice represented by any of the glacial deposits in the bordering region, and even suggest that the changes may have been still more remote.   If the gravel trains in the valleys of the through-flowing streams are, therefore, correlated with the morainic terraces of the Wisconsin age, the history of the drainage of this region would certainly indicate that the drainage modifications produced by the inauguration of the fourth stage of silting must be associated with the earliest known ice invasion, and separated from the Wisconsin by an interval of time represented by the deposition of the silts and the deep erosion of the valley floors, which would represent the whole interval between the earliest ice invasion and the most recent.   Any attempt to determine the length of this interval by comparison with the erosion of the eighth stage, or with the combined erosion of the eighth and ninth stages, either by direct observation of the effects of the two stages in the field or by as careful estimates as are possible to be made in the laboratory, leads to the conviction that this interval is many times greater than any of the other intervals involved in the sequence except those of the older erosion cycles. It thus appears from these studies that they indicate very clearly the general correctness of the interpretation of the glacial geology in the region to the north, and add strength to the view that there was a very great lapse of time between the first ice invasion of this region, which produced the drainage modifications, and the last epoch, which furnished the gravel trains in the deeply cut valleys of through-flowing streams.

If our observations have been correctly made and interpreted, we must conclude that the high-level valleys of southeastern Ohio and the adjacent region represent a connected system of an old drainage cycle which antedates the first advance of the ice of the Glacial period, and that they are therefore of pre-Glacial age; that the deposition of the silts on the old valley floors and the deflection of the streams producing the present drainage system were accomplished as a result of the action of the advancing ice sheet of the first Glacial epoch, possibly the pre-Kansan; that the extensive erosion of the present river valleys to depths below the present drainage lines was accomplished during an interglacial interval of great duration, and that they should, therefore, be called interglacial valleys; that these interglacial valleys were partially filled with débris by the flood waters of the last Glacial epoch and accompanying silting from the adjacent region, with a partial recutting of these deposits during the later stages of this epoch, and that the post-Glacial erosion is represented by the channels cut in the floor of these deposits since the waters have acquired their present volume.

From these studies it appears that the Ohio River Valley, from New Martinsville to Manchester, is of interglacial origin, and that the former considerations, which have been based upon the pre-Glacial age of this valley, are as much in error as those which attempt to attribute its extensive erosion to post-Glacial time.

INDEX.

O

PUBLICATIONS OF UNITED STATES GEOLOGICAL SURVEY.

[Professional Paper No. 13.}

The serial publications of the United States Geological Survey consist of (1) Annual Reports, (2) Monographs, (3) Professional Papers, (4) Bulletins, (5) Mineral Resources, (6) Water-Supply and Irrigation Papers, (7) Topographic Atlas of the United States—folios and separate sheets thereof, (8) Geologic Atlas of the United States—folios thereof. The classes numbered 2, 7, and 8 are sold at cost of publication; the others are distributed free. A circular giving complete lists may be had on application.

The Bulletins, Professional Papers, and Water-Supply Papers treat of a variety of subjects, and the total number issued is large. They have therefore been classified into the following series: A. Economic geology; B, Descriptive geology; C, Systematic geology and paleontology; D, Petrography and mineralogy; E, Chemistry and physics; F, Geography; G, Miscellaneous; H, Forestry; I, Irrigation; J, Water storage; K, Pumping water; L, Quality of water; M, General hydrographic investigations; N, Water power; O, Underground waters; P, Hydrographic progress reports. This paper is the twenty-sixth in Series B, the complete list of which follows. (B=Bulletin, PP=Professional Paper, WS=Water-Supply Paper.)

SERIES B. DESCRIPTIVE GEOLOGY.

B 23. Observations on the junction between the Eastern sandstone and the Keweenaw series on Keweenaw Point, Lake Superior, by R. D. Irving and T. C. Chamberlin. 1885. 124 pp., 17 pls.
B 33. Notes on geology of northern California, by J. S. Diller. 1886. 23 pp. (Out of stock.)
B 39. The upper beaches and deltas of Glacial Lake Agassiz, by Warren Upham. 1887. 84 pp , 1 pl. (Out of stock.)
B 40. Changes in river courses in Washington Territory due to glaciation, by Bailey Willis. 1887. 10 pp., 4 pls. (Out of stock)
B 45. The present condition of knowledge of the geology of Texas, by R. T. Hill. 1887. 94 pp (Out of stock)
B 53. The geology of Nantucket, by N. S. Shaler. 1889. 55 pp., 10 pls.
B 57. A geological reconnaissance in southwestern Kansas, by Robert Hay. 1890. 49 pp., 2 pls.
B 58. The glacial boundary in western Pennsylvania, Ohio, Kentucky, Indiana, and Illinois, by G. F. Wright, with introduction by T. C. Chamberlin. 1890. 112 pp., 8 pls. (Out of stock.)
B 67. The relations of the traps of the Newark system in the New Jersey region, by N. H. Darton. 1890. 82 pp.
B 104 Glaciation of the Yellowstone Valley north of the Park, by W. H Weed. 1893. 41 pp., 4 pls.
B 108 A geological reconnaissance in central Washington, by I. C. Russell. 1893. 108 pp., 12 pls. (Out of stock.)
B 119 A geological reconnaissance in northwest Wyoming, by G. H. Eldridge. 1894. 72 pp., 4 pls.
B 137. The geology of the Fort Riley Military Reservation and vicinity, Kansas, by Robert Hay. 1896. 35 pp., 8 pls
B 144. The moraines of the Missouri Coteau and their attendant deposits, by J. E. Todd. 1896. 71 pp., 21 pls.
B 158. The moraines of southeastern South Dakota and their attendant deposits, by J. E. Todd. 1899. 171 pp., 27 pls
B 159. The geology of eastern Berkshire County, Massachusetts, by B. K. Emerson. 1899. 139 pp., 9 pls.
B 165. Contributions to the geology of Maine, by H. S Williams and H. E Gregory. 1900. 212 pp., 14 pls.
WS 70. Geology and water resources of the Patrick and Goshen Hole quadrangles in eastern Wyoming and western Nebraska, by G I. Adams. 1902 50 pp., 11 pls.
B 199. Geology and water resources of the Snake River Plains of Idaho, by I. C. Russell. 1902 192 pp., 25 pls.
PP 1. Preliminary report on the Ketchikan mining district, Alaska, with an introductory sketch of the geology of southeastern Alaska, by A H. Brooks. 1902. 120 pp , 2 pls
PP 2. Reconnaissance of the northwestern portion of Seward Peninsula, Alaska, by A. J. Collier. 1902. 70 pp., 11 pls
PP 3. The geology and petrography of Crater Lake National Park, by J S. Diller and H. B Patton. 1902. 167 pp., 19 pls
PP 10 Reconnaissance from Fort Hamlin to Kotzebue Sound, Alaska, by way of Dall, Kanuti, Allen, and Kowak rivers, by W. C. Mendenhall. 1902. 68 pp., 10 pls
PP 11. Clays of the United States east of the Mississippi River, by Heinrich Ries. 1903. 298 pp., 9 pls
PP 12. Geology of the Globe copper district, Arizona, by F. L. Ransome 1903. 168 pp , 27 pls.
PP 13. Drainage modifications in southeastern Ohio and adjacent parts of West Virginia and Kentucky, by W. G Tight. 1903. 111 pp , 17 pls.

APRIL, 1903.

[Mount each slip upon a separate card, placing the subject at the top
of the second slip. The name of the series should not be repeated
on the series card, but add the additional numbers, as received, to
the first entry.]

Author.

Tight, W[illiam] G[eorge].

. . . Drainage modifications in southeastern Ohio
and adjacent parts of West Virginia and Kentucky;
by W. G. Tight. Washington, Gov't. print. off.,
1903.

111. III p. 17 pl. incl. maps, 1 fig. 29½ x 23ᶜᵐ. (U. S. Geological
survey. Professional paper no. 13.)
Subject series B. Descriptive geology, 26.

Subject.

Tight, W[illiam] G[eorge].

. . . Drainage modifications in southeastern Ohio
and adjacent parts of West Virginia and Kentucky;
by W. G. Tight. Washington, Gov't print. off.,
1903.

111. III p. 17 pl. incl. maps, 1 fig. 29½ x 23ᶜᵐ. (U. S. Geological
survey. Professional paper no 13.)
Subject series B, Descriptive geology, 26.

Series.

U. S. Geological survey.

Professional papers.

no. 13. Tight, W. G. Drainage modifications in
southeastern Ohio and adjacent parts of
West Virginia and Kentucky. 1903.

Reference.

U. S. Dept. of the Interior.

see also

U. S. Geological survey.

CPSIA information can be obtained
at www.ICGtesting.com
Printed in the USA
BVHW070242020119
536776BV00011B/896/P